Artificial Intelligence for Science and Engineering Applications

Artificial Intelligence (AI) is defined as the simulation of human intelligence through the mimicking of the human brain for analysis, modeling, and decision-making.

Science and engineering problem solving requires modeling of physical phenomena, and humans approach the solution of scientific and engineering problems differently from other problems. *Artificial Intelligence for Science and Engineering Applications* addresses the unique differences in how AI should be developed and used in science and engineering. Through the inclusion of definitions and detailed examples, this book describes the actual and realistic requirements as well as what characteristics must be avoided for correct and successful science and engineering applications of AI.

This book:

- Offers a brief history of AI and covers science and engineering applications
- Explores the modeling of physical phenomena using AI
- Discusses explainable AI (XAI) applications
- Covers the ethics of AI in science and engineering
- Features real-world case studies

Offering a probing view into the unique nature of scientific and engineering exploration, this book will be of interest to generalists and experts looking to expand their understanding of how AI can better tackle and advance technology and developments in scientific and engineering disciplines.

Artificial Intelligence for Science and Engineering Applications

Shahab D. Mohaghegh

CRC Press
Taylor & Francis Group
Boca Raton London New York

CRC Press is an imprint of the
Taylor & Francis Group, an **informa** business

Designed cover image: © Peshkova/Shutterstock

First edition published 2024
by CRC Press
2385 NW Executive Center Drive, Suite 320, Boca Raton FL 33431

and by CRC Press
4 Park Square, Milton Park, Abingdon, Oxon, OX14 4RN

CRC Press is an imprint of Taylor & Francis Group, LLC

© 2024 Shahab D. Mohaghegh

ISBN: 9781032439044 (hbk)
ISBN: 9781032439099 (pbk)
ISBN: 9781003369356 (ebk)

DOI: 10.1201/9781003369356

Typeset in Times
by codeMantra

This book is dedicated to the most important individuals in my life: Narges, Dorna, and Greg.

Contents

Author Bio

Shahab D. Mohaghegh, a pioneer in the application of Artificial Intelligence (AI) and machine learning in the Exploration and Production industry, is a professor of Petroleum and Natural Gas Engineering at West Virginia University (WVU) and the president and CEO of Intelligent Solutions, Inc. (ISI). He is the director of WVU-LEADS (**L**aboratory for **E**ngineering **A**pplication of **D**ata **S**cience).

In addition to more than 30 years of research and development in the petroleum engineering application of AI and machine learning, he has authored four books (*Shale Analytics, Data-Driven Reservoir Modeling, Application of Data-Driven Analytics for the Geological Storage of CO$_2$, Smart Proxy Modeling*) and more than 230 technical papers and carried out more than 60 projects for independents, NOCs (National Oil Company), and IOCs (International Oil Company). He is an SPE (Society of Petroleum Engineering) distinguished lecturer (2007 and 2020) and has been featured four times as a distinguished author in SPE's *Journal of Petroleum Technology* (JPT 2000 and 2005) about AI. He is the founder of SPE's Technical Section, dedicated to AI and machine learning (Petroleum Data-Driven Analytics, 2011). He has been honored by the U.S. Secretary of Energy for his AI-based technical contribution in the aftermath of the Deepwater Horizon (Macondo) incident in the Gulf of Mexico (2011) and was a member of the U.S. Secretary of Energy's Technical Advisory Committee on Unconventional Resources in two administrations (2008–2014). He also represented the United States in the International Standard Organization (ISO) on Carbon Capture and Storage technical committee (2014–2016).

Introduction

Artificial Intelligence will be changing our world even earlier than we used to think. By now, you probably have realized the importance of Artificial Intelligence, which is used and incorporated into many topics and items that you commonly deal with. It is obvious that many people throughout the world are interested in knowing and learning the details of Artificial Intelligence and understanding how AI is used to create interesting characteristics that are exposed to everyone. This book is about Artificial Intelligence for Science and Engineering Applications, which is different from Artificial Intelligence for General Applications. The differences between these two applications of Artificial Intelligence have been briefly mentioned throughout the Internet in the last several years. They are applied as Artificial Engineering Intelligence (AEI) and Artificial General Intelligence (AGI). On the Internet in the past several years, it has been mentioned that these two types of Artificial Intelligence are not the same. It is mentioned that AEI is "Designed to perform engineering tasks" while AGI is "Designed to learn and perform any intellectual task". It is also mentioned that AEI "Can reason and solve problems in a specific way" while AGI "Can reason and solve problems in a general way", and AEI is "Still under development" while AGI is "Still a theoretical concept".

These are interesting topics that are mentioned on the Internet; however, there is another item that has also been mentioned on the Internet that is not correct and will be mentioned in more detail in this book. What seems to be incorrect on the Internet is when it is mentioned that AEI is "Based on mathematical and computational models of engineering principles" while AGI is "Based on knowledge from any field". This definition is now one of the main problems in the way Artificial Intelligence is used for engineering problem-solving, specifically about physical phenomena that will be discussed in detail in this book. This book is about the importance of the correct definition and application of AEI to be used to model and solve physical phenomena, which would be far improved than how modeling and solving physical phenomena was performed in the past century. These are the two reasons why Artificial Intelligence is currently being used to solve multiple different types of science- and engineering-related problems that do not provide the correct results and do not characterize how Artificial Intelligence must be used to comprehensively enhance modeling and solve physical phenomena. It is important to note that solving different types of problems using Artificial Intelligence requires different techniques and approaches.

The first reason has to do with the experts in Artificial Intelligence who tried to simulate, model, and solve science- and engineering-related problems of specific physical phenomena while having no specific expertise on those specific physical phenomena (Petroleum, Chemical, Mechanical, Civil, and any other engineering). The result of such applications of Artificial Intelligence ended up being quite poor

DOI: 10.1201/9781003369356-1

and surprised the industrial management that reached out to original Artificial Intelligence experts to solve their problems.

The second reason has to do with the experts in science and engineering of specific physical phenomena (Petroleum, Chemical, Mechanical, Civil, and any other engineering) who do not have specific expertise in Artificial Intelligence. When such science and engineering experts of physical phenomena try simulation, modeling, and solving science- and engineering-related problems of physical phenomena using Artificial Intelligence, their results also end up having nothing to do with AI. They end up solving such problems in exactly the same way that it was done in the past century, but they now call it AI or Machine Learning results.

The main issue in this book is the requirements of expertise in both "specific science- and engineering-related problems" as well as realistic expertise in "Artificial Intelligence" in order to be able to simulate, model, and solve science- and engineering-related problems of physical phenomena using Artificial Intelligence.

Since mid-2000 (2005–2007), Artificial Intelligence has become interesting and important and has been exposed to general people through the provision of technologies such as image and sound recognition through computers by several companies such as Google, Amazon, IBM, Microsoft, and … many science and engineering industries that have communicated with Artificial Intelligence experts. The management of these industries wanted to test Artificial Intelligence to enhance the results of their solutions. These industries pursued experts in Artificial Intelligence to use the data from their industry's history. Artificial Intelligence experts who were involved in such interactions with the science and engineering industries were experts in the development of AI-based models and systems that are part of General Intelligence such as image recognition, sound recognition, and translation. The results of such interactions between Artificial Intelligence experts (with no specific domain expertise in a specific science and engineering topic) and the science and engineering industries ended up being incredibly different from what the industries expected.

The results of such efforts in the science and engineering industries led their management to conclude that Artificial Intelligence should not be used in the science and engineering industries because it cannot provide any expected reasonable results. In many cases, the management of such industries came to the conclusion that Artificial Intelligence is not a truly good and positive technology. The reality is that the correct application of Artificial Intelligence in modeling and solving science- and engineering-related problems is incredibly better than what industrial management even originally expected and will be covered in this book.

What is today called Artificial Intelligence started as concepts and ideas in the early 1950s. The rule-based "Expert Systems" ended up being referred to as Artificial Intelligence, while the original version of "Perceptron" (data-driven pattern recognition) was moved out of this technology due to a book written by Marvin Minsky and Seymour Papert in 1969 [1]. Some research continued to be done on data-driven pattern recognition, and finally, this technology addressed the issues that were identified by Minsky and Papert in their book and started to be what it is today, in 1986.

During the first 15 years of the 2000s, Artificial Intelligence was extensively used in some games (Chess, Go, …), Internet (Google, Image Recognition, …), Smart Phones, and Tablet computers. In the past couple of decades, the overwhelming majority of the applications of Artificial Intelligence and machine learning that society has been exposed to have been used to address non-engineering (General Intelligence)-related problems. The engineering application of Artificial Intelligence and Machine Learning is quite different from how this technology is used to solve and address non-engineering-related problems that are also known as human-level intelligence, or AGI. Science and engineering domain expertise is an absolute requirement in the science and engineering applications of Artificial Intelligence and Machine Learning.

Unfortunately, recently it has been learned that domain expertise can have both a positive and negative impact on the engineering application of Artificial Intelligence. This will also be discussed in this book. Domain expertise impacts the engineering application of Artificial Intelligence positively when engineers develop a solid understanding of the philosophy, major characteristics, and realistic approach of Artificial Intelligence. On the other hand, domain expertise has been negatively impacting this technology, when the engineers come to the conclusion that only traditional statistical knowledge and understanding of the mathematics behind the Machine Learning algorithms are all that is needed to make them experts in Artificial Intelligence and Machine Learning. Furthermore, many traditional engineers refer to Artificial Intelligence and Machine Learning as a new tool that needs a bit of skill, and they believe that AI is a tool like what Excel used to be and that it's easy for everyone to learn how to use it.

I have spent a long time talking, presenting, and communicating with industry management about this issue. My main communication with them in the past decade was to mention to them that while expertise in Artificial Intelligence is a requirement, it also requires domain expertise in the specific science and engineering topics that are being addressed using Artificial Intelligence. One example of such presentations was at Petro Talk in the Society of Petroleum Engineers [2]. A lot of people in many industries have finally come to the same conclusions. However, unfortunately, while this was (and still is) a fact, agreements with this issue have caused another problem for the Artificial Intelligence application in science and engineering.

After going through a lot of interaction with a large number of science and engineering domain expertise that try to use Artificial Intelligence, including even a very large number of articles that have been published by scientists and engineers that use Artificial Intelligence to solve physics-based problems, the conclusion has been that an incredible number of science and engineering domain expertise that call themselves "Data Scientists" have a very minimal understanding of Artificial Intelligence. The way they use Artificial Intelligence to solve science- and engineering-related problems clearly improves this fact. This is one of the main reasons that this book has been written. Solving physics-based science- and engineering-related problems using Artificial Intelligence requires domain expertise both in science and engineering as well as Artificial Intelligence. It is important to note that becoming an expert in Artificial Intelligence requires as much time, work, and research as has been provided in science and engineering.

Currently, the overwhelming majority of engineers use the term "Machine Learning" instead of "Artificial Intelligence" in the topic of their work and research. In my involvement with their work and research, I always wondered why they did not use the term "Artificial Intelligence". Originally, I was perfectly fine with that, until I recently learned why they do that now. Getting involved with their work and research, I learned that they do research to find out what kinds of Machine Learning algorithms currently exist that they can use to solve their engineering-related problem, which is the topic of their work or research. They search on the Internet and find different versions of Machine Learning Algorithms and use them to solve their problems. They use the data that they have collected in the same fashion that they have solved traditional technologies or the way it has been used in traditional statistics. What they do not pay attention to, is that overwhelming majority of the Machine Learning Algorithms that they try to find on the Internet have been developed to solve AGI items and not engineering-related problems.

They usually end up using several existing Machine Learning algorithms and deciding which one is best for them. Again, the key is that the overwhelming majority of such engineers and scientists who are not experts in Artificial Intelligence just end up using one (or more) existing "Machine Learning" algorithms to deal with the specific engineering-related topic in their work or research. Given the fact that almost all the "Machine Learning" algorithms that they find and end up using have been developed by AI experts, their objective was to solve general intelligence (non-engineering) problems and not engineering-related problems. These engineers end up using "Machine Learning" algorithms for non-engineering-related topics to solve engineering-related problems. This is one of the main problems that currently exist in Science and Engineering Application of Artificial Intelligence.

Again, based on my recent involvement with engineering research projects that include several universities, a large number of national laboratories, and several companies, there is another problem that they are all performing. Beyond using the existing "Machine Learning" algorithms, the next serious problem associated with this type of approach (using existing "Machine Learning" algorithms of General Intelligence to solve engineering problems) is how they use "Data" in their "Machine Learning"-related work and research.

Depending on the type of engineering-related problem that they are dealing with, they usually deal with the data that is usually used to solve that engineering-related problem through our traditional engineering approach, which includes mathematical equations. It should be noted that only using this type of data is quite like traditional statistical approaches. Traditional Statistical approach also uses the data that is usually included in solving engineering-related problems through mathematical equations. Therefore, beyond using general intelligence-related "Machine Learning" algorithms, using similar amounts and types of data that are used in traditional solutions to engineering-related problems seems to be the main reason that they use the term "Machine Learning" rather than "Artificial Intelligence" as the topic of their work and research. This is one of the main reasons that this book is being written: to help all scientists and engineers correctly use Artificial Intelligence and its "Machine Learning" algorithms, as well as the correct way of using "Data" to solve science- and engineering-related problems.

These current approaches provide the conclusion that such scientists and engineers have the idea that Machine Learning algorithms are the new algorithms that must be used in traditional engineering problem-solving as well as traditional statistics. The fact is that this type of approach has absolutely nothing to do with the realistic application of "Artificial Intelligence" in science and engineering. It seems that their main conclusion about this new "Data-Driven" approach only has to do with solving previous problems faster and quicker instead of solving science- and engineering-related problems in a better and more realistic way to come to much better results.

As it is covered in this book, in the next several chapters, it will be understood that the engineering problem solutions using this new "Data-Driven" technology that has been generated through "Artificial Intelligence" are quite different from traditional engineering problem-solving that has been developed more than four centuries ago and Traditional Statistics that has been developed more than one and a half centuries ago. These new approaches have to do with the actual and realistic definition of "Artificial Intelligence" and, based on such a definition, how it must be used to solve science- and engineering-related problems using actual data.

Application of Artificial Intelligence in science- and engineering-related problems is not about using traditional statistics and mathematical equations. To maximize efficiency and spur practical innovations, engineering domain experts must be seriously taught and trained to become experts in Artificial intelligence. The objectives of Science and Engineering Application of Artificial Intelligence include:

- Modeling physical phenomena using facts, reality, and measurements to avoid any assumptions, interpretations, simplifications, preconceived notions, and biases.
- Advancing the art and science of engineering problem-solving, design, and uncertainty quantification with extensive incorporation of Machine Learning algorithms that represent Artificial Intelligence.
- Training the next generation of engineers and scientists with practical knowledge and expertise in the art and science of Artificial Intelligence.

In this book, solving science- and engineering-related problems using Artificial Intelligence is explained, followed by identifying the key requirements of this technology. Furthermore, the reasons behind the lack of success of this technology in many industries will be explained. The following items are absolute requirements and realistic expectations of Science and Engineering Application of Artificial Intelligence:

A. Science and Engineering domain expertise is an absolute requirement for the science and engineering application of Artificial Intelligence,
B. Artificial Intelligence domain expertise is an absolute requirement for the science and engineering application of Artificial Intelligence,
C. The development of comprehensive Artificial Intelligence-related models for science- and engineering-related problems requires *ONLY* actual data (field measurements), and
D. There is no need to include any mathematical formulation (equations) of physics in this process.

It is very important to note that the main reasons for writing this book and the lack of AI-related success in the past several years in many companies are the following:

    i. Inclusion of the term "Physics-based" in their AI-related approach when referring to inclusion of mathematical formulations (equations) of physics,
   ii. They do not understand or have not concluded that the science and engineering applications of Artificial Intelligence are different from the non-engineering application of Artificial Intelligence,
  iii. They do not know what AI-Ethics is, and even if they do, they do not believe AI-Ethics is applicable to engineering problem-solving,
  iv. They have failed to develop comprehensive and explainable Artificial Intelligence for science- and engineering-related problems using ***ONLY*** actual data (field measurements) without the inclusion of mathematical formulations (equations) of physics.

Unfortunately, recently, many engineers started calling themselves "Data Scientists" right after reading a book or a few papers, watching a few "YouTube" videos, or listening to a few lectures. It seems that the reason behind such simplifications is based on two facts:

    1. Scientists and engineers are currently thinking that to handle the physical phenomenon, they need to understand the mathematics behind the modeling of physics, and
    2. The mathematics behind many Machine Learning algorithms that represent Artificial Intelligence are reasonably simple.

Therefore, the reason many scientists and engineers start calling themselves "Data Scientists" so quickly seems to be that once they learn the mathematics behind the Machine Learning algorithms that represent Artificial Intelligence, they come to the conclusion that they know all that needs to be known about the application of Artificial Intelligence in science and engineering. This is the wrong assumption. Such misunderstandings of Artificial Intelligence seem to be the main reason behind the recent development of so-called "Hybrid Models".

The application of Artificial Intelligence is a complete paradigm shift in how science- and engineering-related problems are addressed. Becoming a "Data Science" expert practitioner requires a comprehensive understanding of how to modify the traditional science and engineering problem-solving approach. Understanding the mathematics behind the Machine Learning algorithms that represent Artificial Intelligence contributes less than 15% to becoming a true "Data Scientist", specifically when it comes to science- and engineering-related problem-solving. This is the reason why it is a waste of time to purely concentrate on the mathematics behind Machine Learning algorithms that represent Artificial Intelligence during the short courses that are taught on this topic.

1 Definition of Artificial Intelligence and Machine Learning

In this chapter, the actual definitions of the Artificial Intelligence and Machine Learning will be covered. Unfortunately, many engineers have generated their own definition of Artificial Intelligence and Machine Learning and have not paid attention to the fact that the actual definition of this technology has been defined by the scientists who have originally developed it. It is important to make sure that the personal definitions of such technologies are not used instead of the actual definitions of this technology, which are defined by actual developers and actual experts of this technology. Some engineers generate their own definitions of Artificial Intelligence and Machine Learning so that their unrealistic application of this technology would make sense based on their definitions.

First and foremost, before the definitions, let's go through the real meaning of these four words: "Artificial", "Intelligence", "Machine", and "Learning". Once the actual meaning of these words that are used for this revolutionary technology is described, the actual definition of this revolutionary technology will make much more sense.

WHAT ARE "ARTIFICIAL" AND "INTELLIGENCE"?

First, let's identify the definitions of the two words "Artificial" and "Intelligence". Then putting these definitions together would provide some details about what "Artificial Intelligence" is and why it was used as a term to define this new technology. The word "Artificial" means replicated and is the copy and mimic of something that is "Natural". "Artificial" is made, created, or produced by humans (*Homo sapiens*) rather than occurring naturally. The word "Intelligence" refers to the ability to acquire and apply knowledge and skills. "Intelligence" is the capacity for logic, understanding, self-awareness, learning, emotional knowledge, reasoning, planning, creativity, critical thinking, and problem-solving. Before the term "Artificial Intelligence" anytime the word "Intelligence" was used, it would refer to the Natural "Intelligence" and Human "Intelligence".

Therefore, the term "Artificial Intelligence" should not change the definition of "Intelligence" as what is Natural "Intelligence" and Human "Intelligence". The term "Intelligence" should mean that "*Homo sapiens*" are making and producing what is known as Natural "Intelligence" and Human "Intelligence". In other words, "Artificial Intelligence" is the simulation of Natural "Intelligence" and Human "Intelligence" using machines (computers). Since Natural and Human "Intelligence"

DOI: 10.1201/9781003369356-2

is performed through the human brain, "Artificial Intelligence" mimics the human brain. Therefore, the definition of "Artificial Intelligence" would be the simulation of Natural and Human "Intelligence" through mimicking the human brain using machines (computers).

Now that this definition of "Artificial Intelligence" has been done, it would be interesting to discuss how Natural and Human "Intelligence" works, so that it can be used in the same fashion for the Artificial version of "Intelligence". First and foremost, the major Natural and Human "Intelligence" is "General". The General Intelligence is incorporated into all *Homo sapiens*. The General Intelligence includes thinking, reasoning, remembering, imagining, learning words, using language, and a large amount of knowledge and problem-solving. The key to the definition of General Intelligence is to define Science and Engineering Intelligence. In Human Species, General Intelligence is common, while Science and Engineering Intelligence is not common and requires many studies and learning that goes above and beyond General Intelligence.

Here is a question: Would it be possible for many humans to become expert scientists and engineers without attending high schools and then universities in order to earn their bachelor's degree (B.S.), master's degree (M.S.), or doctoral degree (Ph.D.), and then spend years of working to solve actual physics-based problems and gain experiences? If the answer to this question for human beings is "No", "It is not required", or "It is not possible", would it make sense? Obviously, it would not make any sense since it is an absolute requirement for humans to go through such details in order to become science and engineering experts. Since such issues are quite actual and real about Natural and Human "Intelligence", would it make sense for it not to be similar (actual and real) to "Artificial Intelligence"? In other words, if someone does not have any science or engineering expertise, does it make sense for her/him to be able to use Artificial Intelligence to solve science- and engineering-related problems?

Science and Engineering "Domain Expertise" is an absolute requirement to use "Artificial Intelligence" for solving science- and engineering-related problems. Based on such definitions, it would be obvious that using "Artificial Intelligence" to solve different types of problems requires different types of approaches and solutions, which furthermore require different types of experiences and knowledge since the same is true of Natural and Human "Intelligence". However, given the fact that "Domain Expertise" is an absolute requirement for solving science- and engineering-related problems using "Artificial Intelligence", the same is true about having "Artificial Intelligence" expertise for such solutions. This characterizes the importance of science and engineering domain experts studying "Artificial Intelligence" to be able to solve physics-based problems using "Artificial Intelligence". The important issue that needs to be mentioned here, from a Natural and Human "Intelligence" point of view, is the fact that the same amount of experience, learning, and research for expertise in "Artificial Intelligence" must be done because they have already become domain experts in science and engineering.

It is incorrect to think that being a science and engineering domain expert requires very little detail about "Artificial Intelligence", since such individuals already are scientists and need only to study "Artificial Intelligence" for a few weeks

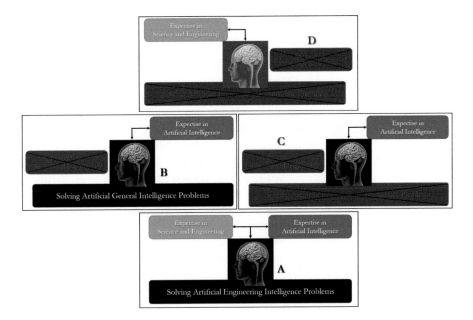

FIGURE 1.1 (a) Expertise in artificial intelligence as well as science and engineering is required for solving engineering-related problems using artificial intelligence, (b) expertise in artificial intelligence is required for artificial general intelligence, (c) only expertise in artificial intelligence is not enough for artificial engineering intelligence, and (d) science and engineering experts that are not experts in artificial intelligence cannot be able to solve artificial engineering intelligence.

or months, different from what they have already done to become experts in their own science and engineering topic. The opposite side of this is just as wrong, when "Artificial Intelligence" experts think that they can solve any kind of problem using this new technology because they are "Artificial Intelligence" experts and need little understanding of any science and engineering problem that they are trying to solve. Figure 1.1 shows (a) the requirement of expertise in Artificial Intelligence as well as expertise in science and engineering to be able to use Artificial Intelligence to solve engineering-related problems (Artificial Engineering Intelligence). It also shows that (b) experts in Artificial Intelligence can solve Artificial General Intelligence but not Artificial Engineering Intelligence (c), and science and engineering experts who are not experts in Artificial Intelligence cannot be able to solve Artificial Engineering Intelligence (d).

WHAT ARE "MACHINE" AND "LEARNING"?

"Machine Learning" is a series of algorithms that are used to generate "Artificial Intelligence". It is a fact that no data-driven problem-solving approaches were referred to as "Machine Learning" prior to the generation and use of "Artificial Intelligence". Since certain types of "Intelligence", such as science and engineering, require

"Learning" and since computers (machines) are the tool for "Artificial" development of "Intelligence" then, the development of "Artificial Intelligence" requires the use of "Machine Learning".

While the details of "Machine Learning" will be discussed, first let's identify the definitions of the two words "Machine" and "Learning". After that, putting these definitions together would provide some details about what "Machine Learning" is and why it was used as a term to define this new technology. The word "Machine" in the context of performing algorithms refers to "Computers". The word "Learning" means the process of acquiring new understanding, knowledge, behaviors, skills, values, attitudes, and preferences. Since "Machine Learning" algorithms are used to develop and generate "Artificial Intelligence", the definition of "Artificial Intelligence" that was mentioned in the previous section makes "Machine Learning" follow the way Natural and Human "Intelligence" go through the learning process. In other words, machines should be exposed to learning the way humans are exposed to learning. However, given the fact that machines (computers) are not the same as humans, while "Learning" has the same types of requirements, it obviously cannot be done in the same fashion.

The idea is to cover how humans learn and then see how computers can learn. In this specific application of "Artificial Intelligence" in Science and Engineering, the human learning of science and engineering should be paid attention to, and then try to find out how machines (computers) can learn the same types of items. In that context, let's first identify: how do humans learn science and engineering? It is quite obvious that for humans to learn science and engineering, they are required to go to high school and university for several years. When humans go to the university to learn about science and engineering, what do they get exposed to too? It is entirely clear that professors at universities "Teach" topics to the students so that they can "Learn" the topics that they are attending in courses. It is obvious that in universities "Teaching" is the best way to "Learning".

In General Intelligence, learning can be done through exposure and experience; nevertheless, teaching can also be helpful, although it may not necessarily be an absolute requirement since "General Intelligence" is part of the brain of the Human Species. However, when it comes to science and engineering, "Learning" absolutely requires "Teaching". Since this is the case in Natural and Human "Intelligence", it must be used for "Artificial Intelligence" through "Machine Learning". As it was mentioned in the previous paragraphs that "Professors at the universities Teach" topics to the students so that the students can "Learn" the topics that they are attending in courses, then it is important to answer the following questions: Is it possible for a professor at a university to "Teach" a topic that she/he is not an expert in? The answer is clear: NO.

In order to teach a topic, you must be an expert in that topic and know it very well so that you can find the best way to communicate and teach the essence and details of the topic to the students that have taken the course so that they can "Learn" it. The same is true about "Machine Learning" in the science and engineering application of "Artificial Intelligence". Domain expertise is an absolute requirement of "Artificial Intelligence" and "Machine Learning" to be used to analyze, model, and solve science- and engineering-related problems. Details of this issue will be covered in more detail in this book.

ARTIFICIAL INTELLIGENCE AND MACHINE LEARNING

Artificial Intelligence and Machine Learning are a complete paradigm shift when compared with traditional approaches to analysis, modeling, problem-solving, and decision-making in science and engineering. Artificial Intelligence and Machine Learning for Science and Engineering Applications is a "Paradigm Shift". Examples of Paradigm Shifts in science and engineering are Copernican Paradigms in the 16th century, Newtonian Paradigms in the 19th century, and Einsteinian Paradigms in the 20th century.

The new Paradigm is analysis, modeling, problem-solving, and decision-making in science and engineering through Artificial Intelligence and Machine Learning in the 21st century. The advancements and breakthroughs in the way the industry is approaching and solving challenging engineering and scientific problems are the paradigm shift in the main issues of Artificial Intelligence and Machine Learning. Artificial Intelligence and Machine Learning, together with the availability of large volumes of data and significant computational capabilities, have rapidly changed the landscape for every aspect of science and engineering.

To start the definition of Artificial Intelligence and Machine Learning, specifically when we are discussing the application of this revolutionary technology in science and engineering, it would be interesting to first read the following two comments and try to find out which one of these two comments makes more sense and is the more correct version of Artificial Intelligence:

> Comment One: Some scientists (original developers of Artificial Intelligence) started this new science and technology (AI) because ***they were interested in analyzing data***. This is how they came up with this technology for data analytics.
>
> Comment Two: Some scientists (original developers of Artificial Intelligence) were interested in the ***creation of intelligence outside of the brain of the Human Species*** and then they learned and concluded that ***this can be done in the best way through dealing with data***.

Even if the first comment (Comment One) does not make you feel too wrong, the actuality of Artificial Intelligence exactly is described in the second comment (Comment Two). Reading the brief history of Artificial Intelligence (in Chapter 2 of this book) will determine why the second comment is all about actuality. Let's start by defining "***Artificial Intelligence***" and "***Machine Learning***".

It must be noted that the human brain is the most powerful pattern-recognition engine in the universe. The human brain includes about 100 billion neurons (also called nerve cells). A neuron, as shown in Figure 1.2, includes several dendrites, a cell body (soma), and an axon that includes several axon terminals. Making an artificial version of the human brain that can be used to generate Artificial Intelligence through Machine Learning

DEFINITION

"Artificial Intelligence" is the simulation of "Human Intelligence" mimicking "Human Brain" for Analysis, Modeling, and Decision-Making.

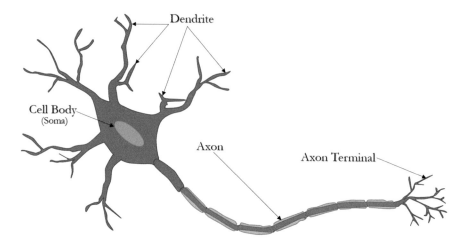

FIGURE 1.2   The neuron in the human brain.

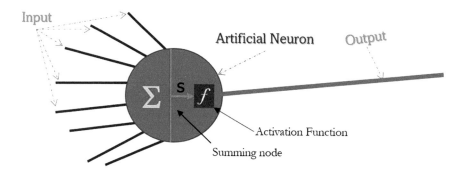

FIGURE 1.3   The artificial version of the neurons of the human brain used in machine learning.

algorithms is shown in Figure 1.3. It should be clear that the artificial version of the human brain is quite similar to what was learned from the neuroscientists, as shown in Figure 1.2.

What is shown as "Dendrite" in Figure 1.2 is shown as "Input" in Figure 1.3. What is shown as "Axon" in Figure 1.2 is shown as "output" in Figure 1.3. In the actual neurons (shown in Figure 1.2), the cell body generates a new type of electrochemical that has been entered into the cell body from the multiple "Dendrites", and this new electrochemical generated by the cell body moves into the "Axon" to be connected to other neurons. In Figure 1.3, the "Artificial Neuron" combines all the "Inputs" (summation of the input values and the weights associated with each input that will represent the quality and importance of each input) and sends it through an "Activation Function" to generate a new value to go to the "Output". While the activities in the cell body of human brain neurons' are bio-chemical activity, in the artificial version of the human brain that is used in machine learning algorithms (Artificial Neural Network) it is a mathematical calculation.

Comparing what is shown in Figures 1.2 and 1.3 clarifies that the scientists that have generated the Artificial Neural Network (Figure 1.3 shows only one artificial neuron) have learned the human brain's characteristics from neuroscience in order to develop the general idea of Artificial Intelligence. Therefore, it is clear that Artificial Intelligence mimics the similar activities of the human brain to learn intelligence and solve problems.

Neurons in the human brain are connected to a large number of other neurons using their axon terminals that are connected to dendrites of many other neurons, Example of such connections are shown in Figure 1.4. The connections between neurons in the human brain are called the "Synaptic Connection", as shown in Figure 1.4. Figure 1.5 shows a more realistic representation of the communication of the neurons in the human brain. The actual number of human brain's neurons is 100 billion. However, the connections of the human brain's neurons to each other develop more than 100 trillion connections between each other. This interaction between neurons in human brain is one of the major characteristics of generating Natural and Intelligence. All the figures in this chapter of the book represent the development of one of the most important Machine Learning Algorithms (Artificial Neural Network) that is commonly used (in so many different ways) to generate Artificial Intelligence to model and solve so many different types of problems.

The interaction between the neurons is done through the "Synaptic Connections", which are mostly about the intimacy and closeness of the axon terminals of one neuron and the dendrites of many other neurons. It is important to note that "Synaptic Connections" in the human brain are not actual physical connections to share

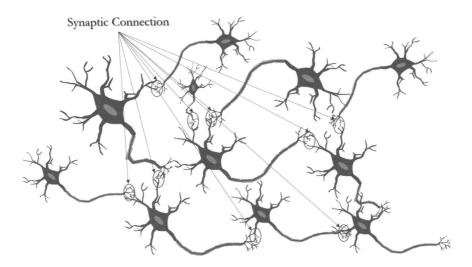

FIGURE 1.4 Neurons in the human brain connect to each other through the axon terminal of one of the neurons and the dendrites of other neurons.

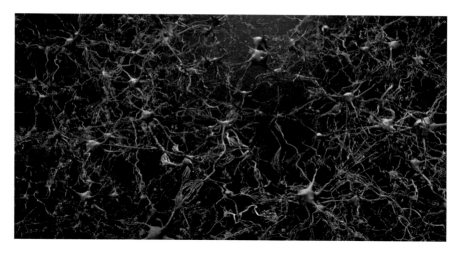

FIGURE 1.5   Example of how neurons in the human brain are connected to each other in large numbers.

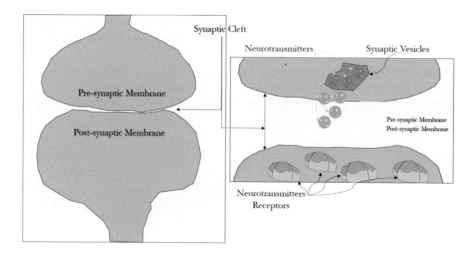

FIGURE 1.6   Synaptic connections between neurons in the human brain.

neurotransmitter molecules, as shown in Figure 1.6. Several neurotransmitters that enter the neurons from quite a lot of other neurons through their Dendrite's electro-chemical interactions enter the cell body.

The cell body creates a new set of neurotransmitters and communicates them with other neurons through its axon. The most common neurotransmitter molecules that communicate between neurons in the human brain are Amino Acids (Glutamate, Gama-Aminobutyric Acid [GABA], and Glycine), Peptides (Opioids: Endorphin, …), Monoamines (Serotonin, Histamine, Dopamine, …), and others (Acetylcholine, …) (Figure 1.6).

Since Human Intelligence is created through the human brain, and since human brain characteristics involve neurons, including their connections and interactions, one of the major issues associated with Artificial Intelligence is mimicking the connections and interactions of the neurons in the human brain. In machine learning algorithms that try to mimic human brains, there are artificial synaptic connections, as shown in Figure 1.7. As shown in this figure, in such Machine Learning algorithms, the outputs of one of the artificial neurons connect to several other artificial neurons through their inputs. In other words, the output of one of the artificial neurons acts as the input of several other artificial neurons. Figure 1.8 provides several numbers of artificial neurons in a Machine Learning Algorithm known as Artificial Neural Network that includes a large number of Artificial Neurons.

Furthermore, it must be noted that another characteristic of the human brain that should also be included in some Machine Learning algorithms has to do with the "Logic" that the human brain uses. It must be noted that in the human brain, logic is also involved in problem-solving and decision-making. One of the main questions that needs to be discussed is: "What kind (type) of logic is used by the human brain?" In the traditional approach to solving science- and engineering-related problems, the logic that is used is mainly "Aristotelian Logic".

Aristotelian Logic that is also used in the current version of computers are two-valued logics. The Two-Valued Aristotelian Logic only includes "0" or "1", "Yes" or "No", "Black" or "White", etc. It is easy to know that the human brain does not use two-valued logic. Instead of only "0" or "1" the human brain includes and deals with infinite numbers between "0" and "1". This is a very well-known fact. This means that Artificial Intelligence and Machine Learning algorithms must use multi-valued logic instead of two-valued logic.

Therefore, the best way to use Artificial Intelligence to model, solve problems, and make decisions would be to reasonably understand how the human brain processes information (data) and what type of logic it uses. This would help the scientists

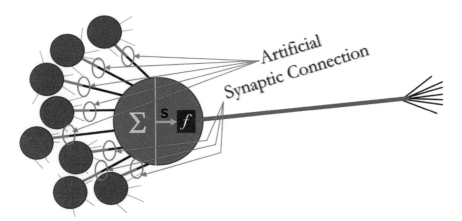

FIGURE 1.7 Artificial neurons in machine learning algorithms are also connected to each other through the output of one of the artificial neurons and the input of other neurons.

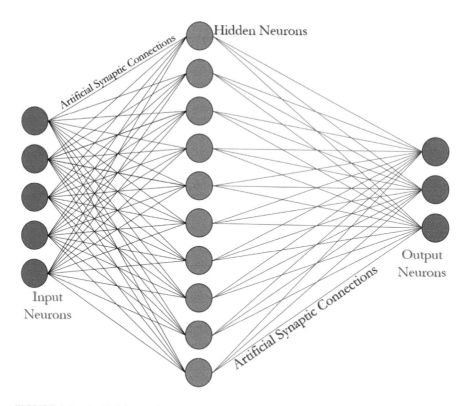

FIGURE 1.8   Artificial neural network machine learning algorithm.

learn what would be the best possible style to mimic the human brain to successfully build and use Artificial Intelligence.

Comparing the speed of actions taken by the human brain versus the chip in your smart phone (or computer), which one is faster? One of the major issues associated with Artificial Intelligence has to do with the minimum amount of time that it takes to solve any problem. Recently, several AI technologies such as ChatGPT, Microsoft Bing, and Google Bard have been developed and provided to everyone through the Internet that are able to answer questions, provide explanations, create writings, language translations, code writing, productivity assistance, text generation, brain-storming ideas, writing different kinds of documents, etc... ChatGPT (Generative Pre-trained Transformer) has been trained through a large dataset of text from the internet that include a wide range of sources, such as websites, books, articles, and other publicly available written material. When this specific question about how it was trained is asked from the ChatGPT itself, one of the items that is mentioned by ChatGPT is: "It is important to note that ChatGPT's responses are generated based on patterns and associations learned from the training data. While the model can pro-vide useful and coherent information, it may also generate incorrect or nonsensical answers at times. It is always advisable to verify information from reliable sources and use critical thinking when interpreting the model's responses".

Google Bard is a Large Language Model (LLM) chatbot[1] developed by Google AI. Google Bard is trained using a large amount of text and code that has been collected from a variety of sources, including Wikipedia, GitHub, and other publicly available data sets. Its training also included datasets from third-party companies and internal data from the company itself. Google Bard is a powerful tool that can be used for a variety of purposes, including education, entertainment, and research.

Another important characteristic of ChatGPT, Microsoft Bing, and Google Bard is the speed at which they perform what has been mentioned. This is also true for all other AI-based models and technologies. The main reason for such a characteristic of Artificial Intelligence has to do with the speed of chips in computers (as well as smart phones, tablets, …). The human brain is 10 million times slower than the computer chips in your smart phone. The speed of neurons firing in the human brain is between 1 and 200 Hz, while the speed of the smart phone CPU is about 2 GHz. Short for gigahertz, GHz is a unit of measurement for alternating current (AC) or electromagnetic (EM) wave frequencies equal to 1 billion Hz. When referring to a computer processor or CPU, GHz is a clock frequency, also known as a clock rate or clock speed, representing a cycle of time. Yet, the human brain has the capability to learn and implement complex activities that is very hard for computers to mimic. This has to do with how the human brain processes information.

Machine Learning includes a series of algorithms that are used to generate Artificial Intelligence. Given the definition of Artificial Intelligence that was mentioned in the previous paragraphs, Machine Learning algorithms are used to process data and use the correct logic. The previous way of using computers (machines) to solve problems was all about explicitly programming the computer and telling it exactly what to do in complete detail. Overwhelming engineers have been using numerical simulation. Developing numerical simulations to solve engineering-related problems that have been modeled using highly complex mathematical equations requires the generation of numerical simulations. All engineers know that during the development of numerical simulation models, they must provide complete, explicit programs so that the computer can perform the numerical simulation. It is important to note that Machine Learning algorithms do not use explicit programming.

The same thing is also true of traditional statistics, which is also a data-driven technology. In traditional statistics and any other traditional type of modeling using computers (machines), we would always tell the computer what to do in exact details without asking the computer to do anything that we had not specifically told it to do.

Before Machine Learning algorithms, we would never expect the computers (machines) to do certain things that we had not told them to do (through coding). This is the main difference between Machine Learning algorithm and what has been done with computers (machines) in the past several decades. Machine Learning algorithms are Open Computer Algorithms. The main issue with Machine Learning algorithms is to let the Open Computer Algorithms learn from the data without using any pre-defined

DEFINITION

"Machine Learning" is the science of getting computers to act without being explicitly programmed through using Open Computer Algorithms to learn from data instead of explicit programming.

mathematical equations that are solved numerically or using a series of mathematical equations to check and see which one of the equations matches the data that is being used. Several general types of Machine Learning algorithms include Supervised Learning, Unsupervised Learning, Semi-supervised Learning, and Reinforcement Learning.

Since the open Machine Learning algorithms will be learning from the data, it would be important to pay attention to the data that will be provided to the Machine Learning algorithms that they are going to learn from. This means that the quality and quantity of the data that is provided to the Machine Learning algorithms are very important. This has two main characteristics that must be paid a lot of attention to:

First, it means that if you intend to include a certain type of "bias" into the Artificial Intelligence modeling using Machine Learning algorithms, then you can do that through the type and amount of data that you would provide to the Machine Learning algorithms.

Second, it means that if you are not a domain expert in a scientific or engineering technology, you will not know what and how to deal with the actual real data that has been collected to provide it to the Machine Learning algorithms so they can learn from the data in a real fashion.

Since "Machine Learning" is a learning process and since the human brain "Learns" the best way on important issues when it is taught in a good and correct manner, "Machine Learning" requires "Teaching". Since we have to provide the correct version of quality and quantity of the data to the machine learning algorithm, we must spend enough time putting together the correct version of data that machine learning will learn from, and depending on the topic that is being modeled, domain expertise plays a critical role in being able to provide the best and correct quality and quantity of the data to the machine learning algorithm.

NOTE

1 A chatbot is a computer program that simulates human conversation through text or voice. There are two main types of chatbots: rule-based chatbots and machine learning chatbots. Rule-based chatbots are programmed with a set of rules that dictate how they should respond to certain prompts or questions. Machine learning chatbots, on the other hand, can learn and adapt their responses based on the conversations they have with users.

2 Brief History of Artificial Intelligence

The general ideas about what is today called Artificial Intelligence started in the early 1950s by an English mathematician, computer scientist, logician, cryptanalyst, philosopher, and theoretical biologist named Alan Turing. His picture is shown in Figure 2.1. He is known as the father of modern computing since he developed the ideas of modern computing and Artificial Intelligence. In one of his original publications [2], he has mentioned, "I propose to consider the question: can machines think?" Alan Turing created a test that has proved a lasting stimulus to later thinkers and contributed to the research of the philosophy and practice of Artificial Intelligence: "Turing Test is a method of inquiry in Artificial Intelligence for determining whether or not a computer is capable of thinking like a human being" [2].

FIGURE 2.1   Alan Turing, the English mathematician, computer scientist, logician, cryptanalyst, philosopher, and theoretical biologist.

DOI: 10.1201/9781003369356-3

John McCarthy   Walter Pitts   Marcian Hoff

FIGURE 2.2   Picture of some of the important technologists who played an important role in the development of Artificial Intelligence.

Pictures of only three of the important technical scientists that played an important role in the development of the Artificial Intelligence in the past several decades (John McCarthy, Walter Pitts, and Marcian Hoff) are shown in Figure 2.2. Since I was not able to provide the actual pictures of many other technical scientists that played an important role in the development of the Artificial Intelligence in the past several decades, in this book, I can only provide their names: Warren McCulloch, Frank Rosenblatt, Bernard Widrow, John von Neumann, Claude Shannon, Allen Newell, Herbert A. Simon, Geoffrey Hinton, Yann LeCun, and Yoshua Bengio.

It is a good idea to use Google Bard and ChatGPT to learn some details about these individuals' contribution to the Artificial Intelligence and several other scientists that have not been mentioned in this book (because of the "Brief History"). What is planned to be covered in this chapter, again briefly, is to simply describe what has happened that the term "Artificial Intelligence" has not been used since 1986, and why, most recently (since the middle of 2005), the term "Artificial Intelligence" started to be used.

ORIGINAL ARTIFICIAL INTELLIGENCE USING DATA AND RULES

It all goes back to the essence of the history of this revolutionary technology (Artificial Intelligence). In the 1950s and 1960s, as the initial version of Artificial Intelligence was generated, both data and rules were used for the development of the Artificial Intelligence models, as shown in Figure 2.3. The data-related models in this original version of Artificial Intelligence were using the initial version of Artificial Neural Networks, which only included two layers of data as "input" and "output".

This original version of the Artificial Neural Networks (called Perceptron) did not include what are today called hidden layers and hidden neurons. Furthermore, the original version of the Artificial Neural Networks would only make its calculations through "Feed Forward" of the neurons, with no technique to move the calculations backward for modification and enhancement of the connections between the input data and the output data. Therefore, the major characteristics of the Artificial Neural Networks (Perceptron) through using data for the model and problem-solving development included "No Hidden Layers", and only "Feed Forward" calculations.

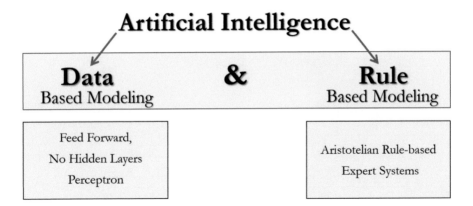

FIGURE 2.3 The initial version of Artificial Intelligence (in the 1950s and 1960s) was developing models based on data and/or rules.

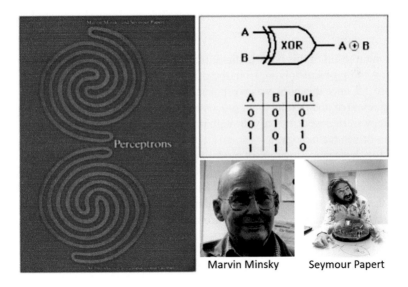

FIGURE 2.4 Marvin Minsky and Seymour Papert wrote an important book *Perceptrons: An Introduction to Computational Geometry* in 1969 that provided a situation to enhance the characteristics of the original artificial neural networks (perceptron).

As shown in Figure 2.4, in 1969, a book named *Perceptrons: An Introduction to Computational Geometry* was written by two university professors from Massachusetts Institute of Technology (MIT), Marvin Minsky and Seymour Papert [1]. In this book, the authors cover the "Perceptron" algorithm (an original version of Artificial Neural Networks) that was generated by an American psychologist named Frank Rosenblatt in 1957. The major contribution of this book to Artificial Intelligence was about the original version of Artificial Neural Networks (Perceptron), explaining that this original version of Artificial Neural Networks can only solve "Linear"-related problems and is not capable of solving any "Non-linear"-related problems.

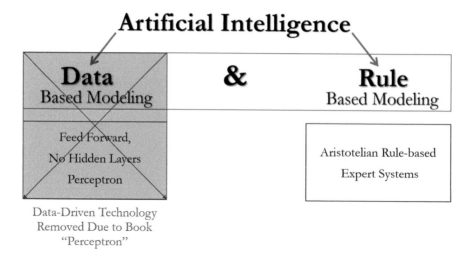

FIGURE 2.5   Data-driven modeling was removed from the original version of artificial intelligence.

The main result of this book is that it literally removes the data-driven version of modeling and problem-solving from the original "Artificial Intelligence". As shown in Figure 2.5, once this book was written and read by a large number of scientists, all existing research and development processes in the existing "Artificial Intelligence" technology started working only through rule-base modeling and avoiding the use of any data-driven technology.

The rule-base modeling type of modeling included a series of rules that must have been followed to solve a specific problem. Such rules would be developed by domain experts about the topic that would be modeled to help humans make decisions. This part of the original version of Artificial Intelligence was called "Expert Systems" and would try to solve complex problems through "If – Then –" rules. One of the major issues associated with this technology was the fact that all the rules that would be generated for "Expert Systems" would be using two-valued, Aristotelian logic.

Developed in the Middle Ages, Aristotelian logic is a classical two-valued logic (either True or False, Yes or No, 0 or 1, …). After the removal of data-driven model development from the original "Artificial Intelligence" for several decades, rule-based "Expert Systems" became the sole focus of research and development in the field of original "Artificial intelligence". Being only based on rule-based "Expert Systems", original "Artificial intelligence" ended up being quite limited in solution development, and no major issues or problems were solved in such a fashion that a large number of people would be exposed to it and find it interesting to work with this technology.

In this context, the original version of Artificial Intelligence that was purely based on rule-based "Expert Systems" ended up not being a new technical revolution, being highly used, and having the capability to solve many problems that could not have been solved in any other way. From this point of view, the original version of Artificial Intelligence was not a great and interesting topic. This was the main reason behind the fact that when in 1986 the new Artificial Neural Network, such as Machine Learning algorithms, were developed and became able to solve both linear and non-linear related problems, the enhancement of this original technology was no longer called "Artificial Intelligence".

These new series of Machine Learning algorithms that were enhancing the original "Artificial Intelligence" and making it a data-driven technology ended up having multiple different names, such as "Computational Intelligence", "Virtual Intelligence", "Soft Computing", "Intelligent Systems", … The key was that because the term "Artificial Intelligence" was not generating and developing any positive and enhanced technologies and interesting problem-solving due to the use of rule-based "Expert Systems", it would no longer be used as the term for data-driven model development and major solutions.

In 1986, the original version of the Artificial Neural Networks that was proven not to be able to solve "non-linear" problems was changed through an algorithm called "Back Propagation". This new algorithm changed all the characteristics of the original version of the Artificial Neural Networks (Perceptron). First and foremost, the new version of Artificial Neural Networks would not only have two layers, such as "Input" and "Output". As presented in Figure 2.6, this new version of Artificial Neural Networks would include at least one or even more layers of neurons between the two "Input" and "Output" layers. This new layer (Layers) was called the "Hidden Layer" since it would not require any data from the "Input" or "Output" layers. This new "Hidden Layer" that was included in the new version of the Artificial Neural Networks would include several hidden neurons that were usually larger than the number of input characteristics. These large numbers of hidden neurons inside the "Hidden Layer" would be connected to all the data and characteristics in both the "Input" and "Output" layers.

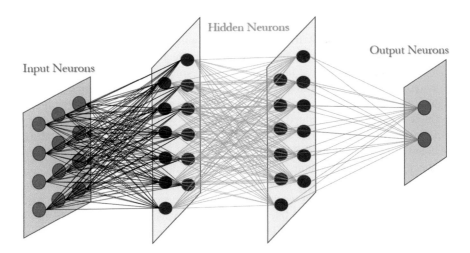

FIGURE 2.6 New version of the artificial neural network that includes multiple layers of neurons between input and output and backpropagation to change the weights between neurons.

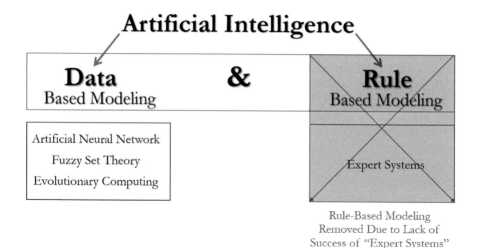

FIGURE 2.7   Rule-based system was removed from the new version of Artificial Intelligence starting in the mid-1980s.

Furthermore, instead of being only a "Feed Forward" process, the new version of the Artificial Neural Networks will also include a "Feed Backward" process through "Back Propagation" algorithms in order to modify and enhance the connections between all the neurons in the multiple layers. In addition to the new version of the Artificial Neural Networks that are currently called machine Learning Algorithms for Artificial Intelligence, other machine learning algorithms are also part of Artificial Intelligence, such as Fuzzy Set Theory and Evolutionary Computing. As shown in Figure 2.7, the current version of Artificial Intelligence no longer includes rule-based "Expert Systems".

However, it is quite possible that in the near future, the current revolutionary technology that is called Artificial Intelligence will use "rule-based" technologies in addition to and in combination with data-driven technologies in its model developments. However, given the fact that the new revolutionary Artificial Intelligence technology simulates natural intelligence and mimics the human brain, it needs to follow the way the human brain uses logic. As it is a well-known fact, the human brain does not use a two-valued logic in its process. When we are asked, "How Are You?" we never only use either the words "good" or "bad" but rather we respond such as "I am doing fine", "I am doing OK", or "I am not so well now", ... Instead of using only two-valued logic, the human brain uses multi-valued logic.

Therefore, instead of using two-valued logic, if "rule-based" modeling, decision-making, and problem-solving would use Fuzzy Set Theory to use multi-valued logic, then the revolutionary Artificial Intelligence technology can use rule-based algorithms in combination with other machine learning algorithms in order to advance the analysis, modeling, decision-making, and problem-solving of science and engineering. As it is shown in Figure 2.8, it is possible (not necessarily for sure) that Artificial Intelligence can go back to its previous shape of using both data-based modeling and rule-based modeling to model and solve problems, with the difference of using multi-valued logic rather than simplified "two-valued" logic.

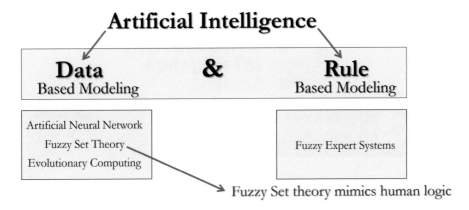

FIGURE 2.8   The future version of Artificial Intelligence most probably would also include rule-based systems, but instead of two-valued logic, it most probably would use multiple-valued logic (Fuzzy Set Theory) like the human brain.

USING THE TERM "ARTIFICIAL INTELLIGENCE"

As was mentioned, starting in 1986, the backpropagation algorithm started to be used in Artificial Neural Networks. This created a large amount of interest in using this machine learning algorithm to solve problems. The overwhelming majority of such applications were done at the Institute of Electrical and Electronic Engineering (IEEE). However, since Artificial Intelligence had become purely a rule-based technology and its characteristics had not created a serious amount of problem-solving interest in scientists, no one used the term "Artificial Intelligence" when this new version of Artificial Neural Networks was being used to solve problems.

In the IEEE, instead of Artificial Intelligence this technology was named "Computational Intelligence", and in many other cases, several other names were used instead of "Artificial Intelligence". For example, in the monthly journal of the *Society of Petroleum Engineering* (SPE), the *Journal of Petroleum Technology* (JPT), three articles was written in the year 2000 about Artificial Neural Networks [3], Evolutionary Computing [4], and Fuzzy Logic [5]. In all these three articles, we referred to "Virtual Intelligence". Other names that were referred to instead of "Artificial Intelligence" included "Soft Computing".

Finally, in the mid-2000s, when Deep Learning and Convolutional Neural Networks were used for image recognition in a fashion that was exposed to general people throughout the world by Google, the term "Artificial Intelligence" started to be used instead of "Computational Intelligence", "Virtual Intelligence", "Soft Computing", etc.

Familiarity with the physics-based and non-physics-based history of Artificial Intelligence and Machine Learning provides some important perspective for those who are interested in the application of this technology in science- and engineering-related disciplines. For example, it is interesting to know that from the late 1980s until the end of the 1990s, many of the scientists and innovators who had initiated the application of this technology in many science- and engineering-related areas intentionally avoided using the term "Artificial Intelligence" to describe their research and development efforts[1].

FIGURE 2.9 IEEE's computational intelligence website.

During the 1990s, applications of Machine Learning algorithms such as Artificial Neural Networks, Fuzzy Set Theory, and Evolutionary Computations in science and engineering were referred to as "Computational Intelligence", "Virtual Intelligence", "Soft Computing", and a few other names. In the mid-2000s, when image recognition technology was generated using a combination of traditional statistics and multi-layer Artificial Neural Networks, the name "Artificial Intelligence" started to be used far more often. At this point in time, the term "Artificial Intelligence" is the main name of the technology that is mentioned when Machine Learning algorithms are used to model or solve problems. It would be interesting to know why the use of the name "Artificial Intelligence" was being avoided for about two decades, from the mid-1980s to the mid-2000s.

There were enough reasons for such avoidance that they even influenced IEEE decision-making in this area. IEEE is the original homeland of the overwhelming set of technologies that were developed and are now contributing to Artificial Intelligence and Machine Learning. When IEEE decided to form a scientific section (council/society) for this technology, it modified the "IEEE Neural Network Council" that was formed in 1989 and later in 2001 was changed to "IEEE Neural Network Society" and started calling it "IEEE Computational Intelligence Society" in 2003, not "IEEE Artificial Intelligence Society" (Figure 2.9).

The main reason the term "Artificial Intelligence" was being avoided by the scientific community at the time had to do with multiple factors. However, the most important fact was the approach to problem-solving at the start of this technology in the 1960s. At the time, "Artificial Intelligence" was completely based on Aristotelian logic rules instead of data-driven approaches. This "rule-based" version of "Artificial Intelligence" that was also used to be called "Expert Systems" failed to achieve any of the large promises that were made, resulting in the waste of extensive amounts of investment that were made in this technology. The final outcome was the failure and collapse of this version of "Artificial Intelligence" (Expert System).

NOTE

1 My first paper on this topic was published in August 1994 (SPE 28237). Just like other (non-petroleum engineering) colleagues in this community, I avoided using the term Artificial Intelligence in my first 29 related publications on this topic prior to the year 2004. The first time I used the term "Artificial Intelligence" was in my 30th paper on this topic, which was published in the JPT (Journal of Petroleum Technology) in May 2004.

3 Science and Engineering Application of Artificial Intelligence

Since "Artificial Intelligence" is the simulation of "Human Intelligence" through "Mimicking Human Brain", to explain the science and engineering application of Artificial Intelligence, it makes sense to first characterize how science and engineering are applied to Human Intelligence. Once the application and characteristics of science and engineering in Human Intelligence are well understood, it would make sense to identify how science and engineering in Artificial Intelligence should be correctly applied since they should be simulating science and engineering on Human Intelligence. As some companies, faculties, and individuals (including scientists and engineers) come to the conclusion that science and engineering applications in Artificial Intelligence have nothing to do with science and engineering applications in Human Intelligence, it becomes quite clear that their understanding of Artificial Intelligence would be very minimal and may even be incorrect.

In this chapter, General Intelligence and Science and Engineering Intelligence of *Homo sapiens* will be covered, followed by an explanation of how the same types of intelligence (general as well as science and engineering) should be simulated using Artificial Intelligence. It seems quite understandable that there are significant differences between General Intelligence and Science and Engineering Intelligence for *Homo sapiens*. Of course, it is obvious that without General Intelligence, it is impossible to create and generate science and engineering intelligence in human species. Since the history of evolution clarifies the importance and uniqueness of *Homo sapiens*' General Intelligence, it completely makes sense that Science and Engineering Intelligence can become part of *Homo sapiens* Intelligence.

HOMO SAPIENS' GENERAL INTELLIGENCE

General Intelligence in humans starts from the day that we are born. General Intelligence in humans is continuously enhanced. Items such as Image Recognition, Sound Recognition, Listening, Speaking, Language Translation, … are achieved through humans' General Intelligence in a reasonably short period of time. Humans' General Intelligence makes even quite young kids able to solve certain types of problems. For example, as shown in Figure 3.1, even a 5-year-old child can tell which one of these images is showing the picture of a cat and which one is showing the picture of a dog. General Intelligence of humans can solve such problems through regular learning and understanding, even during our early lives. A little child can

DOI: 10.1201/9781003369356-4

FIGURE 3.1   Image recognition. Difference between cat and dog images.

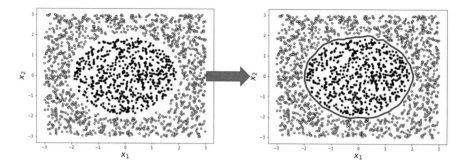

FIGURE 3.2   Black and yellow-colored circles can be separated by a non-linear line.

soon learn which one is "Mom" and which one is "Dad". At an incredibly early stage of her/his life, the child can tell if "Mom" is saying something or "Dad" is talking. In other words, Image Recognition and Sound Recognition are part of our General Intelligence, which happened to us at an incredibly early time in our lives since our General Intelligence was used.

Another example is shown in Figure 3.2. Even before a child goes to school at the age of 7, if you show them a series of black and yellow circles and ask them if they can use a line to separate them from one another, the possibility of the child being able to do it is very high. Almost 99.99% of the children will be able to draw the red line that is shown in Figure 3.2. Examples of Figures 3.1 and 3.2 show the fact that human General Intelligence, even in a short time of life, allows us to solve a lot of problems. Currently, we are using Artificial Intelligence to solve these same types of problems incredibly quickly. Prior to the current version of Artificial Intelligence, solving a problem as shown in Figure 3.1 could not have been done easily and fast by computers, as it is done today using Artificial Intelligence using Machine Learning algorithm called the Convolutional Neural Network (CNN).

It is also important to note that when recently asking ChatGPT and Google Bard about Artificial Intelligence applications in general and engineering-related problem modeling and solving, they mentioned: "General Intelligence aims to replicate human-like intelligence across various domains, while engineering intelligence focuses on building specialized Artificial Intelligence Systems for specific tasks or applications. General Intelligence represents a more comprehensive and broader goal, whereas Engineering Intelligence deals with practical and targeted implementations of Artificial intelligence in specific areas. It is important to note that these are just two different approaches to Artificial Intelligence. There are many other ways to define and categorize Artificial Intelligence. However, general and engineering intelligence are two of the most common approaches".

General Intelligence eventually covers a human's entire life and has particularly important and complex characteristics. I strongly recommend covering the important history of *Homo sapiens* by reading the book *Sapiens: A Brief History of Humankind*, written by Yuval Noah Harari [6]. Human's General Intelligence covered our history for almost 100,000 years before Science and Engineering Intelligence got started in our history almost 300 years ago. Prior to *Homo sapiens*, other human species such as Neanderthals, *Homo erectus*, and … had certain amounts of General Intelligence (not the same amount and type that *Homo sapiens* had) that never changed or were enhanced to Science and Engineering Intelligence. However, it did happen to *Homo sapiens* and changed everything in our history.

To cover *Homo sapiens'* General Intelligence, it is important to cover and understand what happened in human history. Based on a book written by Charles Darwin [7], it is probable that "Africa was the cradle of humans because our two closest living relatives (Chimpanzees and Gorillas) live there" (Figure 3.3). While *Homo erectus* left East Africa almost a million years ago and Neanderthals left East Africa almost 400,000 years ago, they never changed from being hunter-gatherers. Figure 3.4 shows the historical movements of *Homo erectus*, Homo Neanderthalensis (Neandertals), and *Homo sapiens* from East Africa to the rest of the world.

Homo sapiens entered from Eastern Africa into what is today called the "Middle East" and "Europe" about 70,000 years ago. They lived for about 58,000 years as "Hunter-Gatherers". The "Agricultural Revolution" took place about 12,000 years ago. It was a major change in *Homo sapiens'* ways of living when compared to Neandertals

FIGURE 3.3   Human evolution from chimpanzees and gorillas.

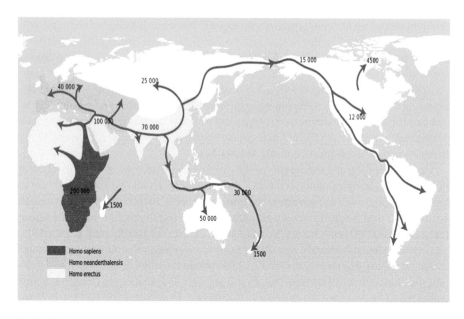

FIGURE 3.4   Historical movement of *Homo erectus*, Homo neanderthalensis (Neandertals) and *Homo sapiens* from East Africa to the rest of the world.

and *Homo erectus* who had never changed their "Hunter-Gatherer" processes for more than hundreds of thousands of years. This demonstrates the enhancement of human General Intelligence.

The next revolution for us (*Homo sapiens*) took place more than 11,000 years after the "Agricultural Revolution". "Industrial Revolution" that happened in the 18th century changed much more issues in people lives (Figure 3.5). It created industrialization, manufacturing, economic transformation, the use of iron, steel, coal, electricity, etc. The objective was to make it possible to perform many actions much faster, easier, and quicker than they could have possibly been done by humans prior to this revolution. This must have had to do with the *Homo sapiens*' brain and its General Intelligence.

It took the industrial revolution about two centuries to completely control and change the evolution of societies. Such revolutions are the major determinations of the difference between the intelligence of *Homo sapiens* and the intelligence of Neandertals and *Homo erectus*. It is completely clarified that *Homo sapiens*' General Intelligence is quite strong and is one of the main reasons why *Homo sapiens* became able to look so different not only from animals but also from all other human species. The main characteristic of such intelligence is the human brain. While the "Industrial Revolution" simulated human muscles, the current advancement in our world is known as the "biotech and Artificial Intelligence" revolution, which would simulate the human brain. This revolution does not require centuries, but only decades, to change our societies.

Artificial Intelligence that simulates human intelligence and mimics the human brain is part of a new revolution in *Homo sapiens* that will change our world in the next few decades (Figure 3.6). Of course, this new revolution in our world is a combination of Artificial Intelligence and Bio-Tech. The Bio-Tech mentioned here refers

FIGURE 3.5 Industrial revolution of *Homo sapiens* 300 years ago.

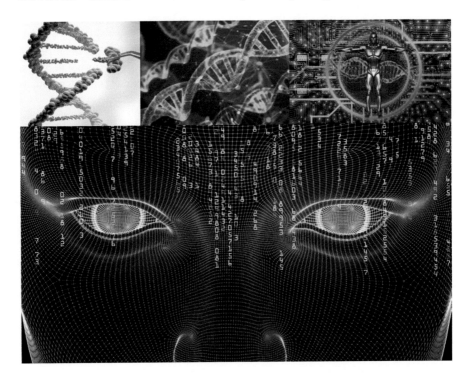

FIGURE 3.6 *Homo sapiens* new revolution: Artificial Intelligence and Bio-Tech.

to CRISPR (Clustered Regularly Interspaced Short Palindromic Repeats) that was developed in 1987 and CRISPER Cas9 (CRISPR-associated protein 9) that had to do with human genes that were generated in 2002 [8].

Artificial Intelligence has already started to change so many things. Science and engineering domain experts that become highly skilled Artificial Intelligence practitioners are the ones that will control the future of science and engineering disciplines. Becoming a science- and engineering-related "Artificial Intelligence expert practitioner" requires extensive experience using Artificial Intelligence to solve science- and engineering-related problems.

It is important to note that becoming a scientist or engineer with expertise in Artificial Intelligence will not happen in an abbreviated period of time. It requires a reasonable amount of time to practice Artificial Intelligence in a realistic fashion, rather than using it as a marketing tool.

HOMO SAPIENS' SCIENCE AND ENGINEERING INTELLIGENCE

In this chapter, in Figure 3.2, it was shown that the non-linear separation of two colors of black and yellow circles can be done by General Intelligence even by young children who may not have been to school or have not yet learned any mathematics. However, if the same young children (individuals) are asked to perform a "Linear Separation" of the two colors of black and yellow circles instead of a circle (non-linear separation), they will not be able to do it since they are not familiar with mathematical equations that can make the black and yellow circles be able to be separated by a linear line, as shown in Figure 3.7.

Today, Artificial Intelligence is a technology that almost everyone is interested in. This technology has proven to accomplish so much while being incorporated in so many areas, such as social media, online search, the stock market, the interaction of

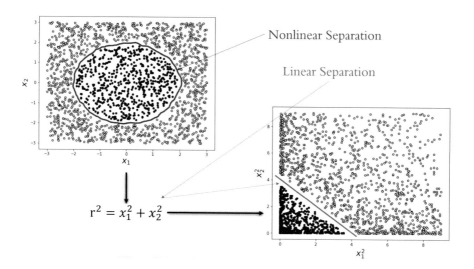

FIGURE 3.7   Linear separation of black and yellow circles using a mathematical equation.

companies with their individual clients, such as Netflix, banks, insurance companies, travel agencies, and even security. Therefore, it makes sense when engineers express interest in incorporating this technology into their everyday work. Given the fact that science and engineering applications of Artificial Intelligence are not exactly the same as the application of this technology to non-engineering-related problems, there have been some issues associated with how some scientists and engineers in academia and industry have been using this technology to solve engineering-related problems. The next item that must be noted is answering the following questions. The answers that are given to these three simple questions should be quite easy and very honest.

QUESTION 1

Is there a difference between General Intelligence and Science and Engineering Intelligence?

ANSWER 1

Yes, Indeed.

QUESTION 2

What is the reason for answering "Yes" to the above question?

ANSWER 2

Because General Intelligence is included in every *Homo sapiens*, those who like to have Science and Engineering Intelligence must go through important learning.

QUESTION 3

How would Science and Engineering Intelligence learning happen in *Homo sapiens*?

ANSWER 3

To generate the Science and Engineering Intelligence *Homo sapiens* must go to high school and university and must at least receive a bachelor's degree.

These questions and answers should be very simple and clear. No one can become a scientist or an engineer without going to university and receiving a bachelor's, master's, or PhD. After completing the university, they will go to industries and academia and work on these topics for years. These are the items that make them scientists and engineers and increase their intelligence from General Intelligence to Science and Engineering Intelligence, which already includes the General Intelligence as well.

In order to explain what needs to be done and what needs to be avoided when Artificial Intelligence and Machine Learning are used to solve science- and engineering-related problems, let us start with the names of these technologies: *"Artificial Intelligence"* and "*Machine Learning*". "*Artificial Intelligence*" has to do with "*Intelligence*" that is not "Natural" but rather "Artificial", and "*Machine Learning*" has to do with "*Learning*" that is applied to "Machines" and computers, not to "Humans". Therefore, the two words to be defined and discussed are "*Intelligence*" and "*Learning*".

"*Intelligence*" is defined as the ability to acquire, retain, and apply knowledge; learn from experience; adapt to new situations; and solve problems and "*Learning*" is defined as the process of acquiring new understanding, knowledge, and skills. After defining these two items, let's start by asking two simple questions about engineering problem-solving:

1. Can human-level General "*Intelligence*" be used to solve science- and engineering-related problems?
2. Do humans need to "*Learn*" anything specific to be able to solve science- and engineering-related problems?

The answers to the above two questions are quite obvious. To become a scientist or engineer, humans need to go to university after they get their high school diploma. Then, after 4 years of "*Learning*" at the university, they become scientists or engineers with bachelor's degree. Furthermore, "*Learning*" at the university goes beyond just providing information to the students through books, videos, and articles. Science and engineering students go through an extensive learning process in 4 years for bachelors, 6 years for bachelors and masters, and 10 years for bachelors, masters, and Ph.D.s, before becoming scientists or engineers. It is also important to note that "*Learning*" at the university includes "*Teaching*". These simple answers to the questions regarding "*Intelligence*" and "*Learning*" mean that the application of Artificial Intelligence in science- and engineering-related problem-solving (*a*) requires science and engineering knowledge and domain expertise, and (*b*) requires "Teaching" through detailed understanding and communication of the data to the machine learning algorithms.

This goes way beyond just throwing the available data to the Artificial Intelligence and Machine Learning algorithms to solve problems, as is mostly done when people use traditional statistics for data analysis. Normally, when traditional statistics is used to solve problems, all that matters is "Correlations", and it usually has nothing to do with "Causation" (cause and effect relationships when modifications in one of the variables produce modifications in another variable). On the other hand, all scientists and engineers are incredibly interested in understanding "Causation", above and beyond just "Correlations". This is what is being called "Explainable AI – XAI". The science and engineering application of Artificial Intelligence and Machine Learning requires "Explainable AI – XAI" in order to be able to generate details about how and why certain predictions are made, beyond just "Correlations". This topic "Explainable Artificial Intelligence" is covered in Chapter 7 of this book.

When the Science and Engineering natural Intelligence of humans requires a significant amount of learning above natural General Intelligence, does it make sense that the same thing would be applicable to Artificial Intelligence? It means that there are differences between *Artificial* Science and Engineering Intelligence and *Artificial* General Intelligence in the same fashion that there are differences between

Natural Science and Engineering Intelligence and *Natural* General Intelligence. This must be obvious. However, unfortunately a lot of people that are interested in Artificial Intelligence do not realize the importance of such differences. Such differences between *Artificial* Science and Engineering Intelligence and *Artificial* General Intelligence provide characteristics on how they should be generated and used to solve science- and engineering-related problems versus non-science- and non-engineering-related problems using *Artificial* Intelligence.

It is important to note that what is mentioned about these differences does not mean that the machines must also go to high school and university and get at least a bachelor's degree in order to be able to solve science- and engineering-related problems, since such an issue is impossible. Therefore, if that is not the case, then how Artificial Science and Engineering Intelligence can be developed to produce correct, interesting, and important results? First and foremost, it must be noted that for a long time (even now in many cases), many industries have Artificial Intelligence experts with no domain expertise in any specific science and engineering to perform Artificial Intelligence which means solving science- and engineering-related problems. This has happened many times in many industries (and in some universities) and the results have proven that it is not a good idea at all.

Here is an example that happened in the petroleum industry [9]. On March 31, 2014, Hitachi provided an online News Releases[1] with the title "Hitachi and EERC proposed Innovative Solutions for Optimizing – the Bakken". Then it was mentioned that "Data Management for the one of the Largest Oil Field in the United States".

The start of this online News Releases mentioned:

"Hitachi America, Ltd and the Energy & Environmental Research Center (EERC)" today announced that the two companies are partnering to develop technology-driven products, services, and solutions for enhancing the production of oil reserves from the Bakken formation. Covering parts of North Dakota, Montana, Saskatchewan, and Manitoba, the Bakken formation is estimated to hold hundreds of billions of barrels of oil".

"With nearly 6,000 unconventional Bakken petroleum system wells and nearly 200 drilling rigs in North Dakota alone, the sheer volume of data collected to date is staggering. Using powerful analytical tools to compile and evaluate Bakken data, oil-field operators and state officials can better understand the relationships between the productivity of an oil well and a variety of factors, such as how the well was drilled and the geology of the area in which it was drilled. These tools can also be used to identify previously unrecognized relationships in the data. In turn, this improved understanding will enable industry to improve the productivity of current and future Bakken wells".

Once EERC provided Hitachi all the data, apparently Hitachi did not have or did not use a petroleum engineering domain expert who is also an AI expert to perform this project for EERC. The result that Hitachi concluded from the data that was provided to them was that oil production from Bakken (North Dakota, Montana, Saskatchewan, and Manitoba) was a function of weather rather than anything about petroleum engineering. Coming up with such solutions using "Data Science" without any expertise or understanding of petroleum engineering makes perfect sense.

The engineers and scientists in the EERC were shocked by such results from Artificial Intelligence and communicated with Hitachi about how incorrect such results can be.

The response of Hitachi to EERC's engineers and scientists was also published on Friday, April 24, 2015, with the topic "UPDATE! Data Blending, Spurious Correlations and Rainy Day". In this article, it was mentioned:

"It has come to my attention that the result from this analysis presented in this blog is overstated in its conclusions. The folks at the EERC have pointed out two areas of discrepancies that I would like to address here, the volume of production shown in the charts below is overstated (the y-axis) and the amount of cold weather production decline is overstated (the x-axis). After examining these details with Matthew O'Keefe, it turns out they are right. You've got to love experts!"

"First, the volume of production shown on the y-axis is wrong. Operationally, the data used for this calculation boils down to an operational error on my part – the database was loaded more than once with the same data resulting up to three copies of the same data existing in this table. This greatly inflated any results in the analysis".

"This has an impact on the second factor of cold weather's impact on production. In the same chart, an "up to an 80% reduction impact" was stated during periods of extreme cold weather months. This is in fact wrong by a significant factor. The decrease in production is closer to 9%–15% from peak production to lowest production in recent years, as this new chart shows here" and also "How does weather impact oil production in the Bakken?" "I will need some weather data, specifically historical weather data for the region". The oil production data for the region goes back to 1960, so there is a lot of weather history to work with. Using the National Oceanic and Atmospheric Administration – NOAA website (http://noaa.gov), I was able to obtain historical weather data from the Bakken region, using a latitude and longitude polygon, since 1960, the first oil production entry in my oil production dataset". Around the mid-2000, when Image Recognition was developed by Artificial Intelligence and Machine Learning and demonstrated its strong value and characteristics through data, the management of one of the most well-known petroleum companies in the United States drew a series of Artificial Intelligence experts (without any petroleum engineering expertise) and asked them to use their company's incredible amount of data to provide them with the opportunity to enhance oil and gas production.

When the Artificial Intelligence experts performed their analysis using the data that was provided to them, they mentioned the following to the petroleum companies' management: "In order to enhance your oil and gas production, you need more porosity in the reservoirs that you work on". Since this information is taught to the petroleum engineering students in their first reservoir engineering courses, and therefore it is a well-known fact for people that start learning about petroleum engineering, the petroleum companies' management was shocked that such a simple petroleum engineering technology was the result of the Artificial Intelligence experts. Such issues were the reason that many industries (during the early stages of the application of Artificial Intelligence in their companies) concluded that Artificial Intelligence does not work or is a waste of time and money. What they did not realize was that such problematic issues and results are not the problem of Artificial Intelligence; it is about not using it correctly.

PROBLEMS ASSOCIATED WITH THE SCIENCE AND ENGINEERING APPLICATION OF ARTIFICIAL INTELLIGENCE

Currently, the main issues associated with the science and engineering application of Artificial Intelligence and Machine Learning in academia and industry have to do with three major problems: (1) When Artificial Intelligence or statistics experts that have a minimal understanding of engineering are identified as the main leaders and scientists to address and solve engineering-related problems using data; (2) The involvement of traditional engineers that do not believe Artificial Intelligence is science and technology and claim that Artificial Intelligence is a hoax (same as Covid-19) and should not be used to solve engineering-related problems; and (3) The engineers who have a minimal and superficial understanding of Artificial Intelligence and are mostly interested in it from a business and marketing point of view rather than science and technology.

Here are simple explanations of each of the three problems that have been mentioned:

First Problem: **When Artificial Intelligence or statistics experts with minimal understanding of engineering are identified as the main leaders and scientists to address and solve engineering-related problems using data.**

One of the major problems that many companies in several industries started with, and some still are doing it, is hiring AI experts and/or statisticians to apply Artificial Intelligence and Machine Learning to their industry. In 2018, in a Petro-Talk [10], I explained the main problems with this approach in the petroleum industry. Those who do not have engineering domain expertise usually use the data that is provided to them and try to find "Correlations" between the parameters that are shared with them. They do not pay much attention to finding out whether the data that is provided to them includes all that is required to accomplish the objective that is being addressed. Or deal with the provided data in a fashion that would teach the machine learning algorithm about the characteristics of the physical technology that is being modeled.

Lack of domain expertise does not allow them to identify what data is needed in order to solve the problem that they are dealing with. Is it possible for a professor who has no knowledge about a specific engineering technology to teach that specific engineering technology-related course to the students? In other words, can a political science professor teach thermodynamics? Or vice versa? As far as natural "*Intelligence*" is concerned, dealing with engineering-related problems requires engineering domain expertise. Therefore, the same would be true when it comes to artificial "*Intelligence*". A domain expert engineer who becomes an expert practitioner of Artificial Intelligence will be able to generate the required features that can be used to "Teach" the Artificial Intelligence and Machine Learning algorithms how to solve specific engineering-related problems.

Becoming an expert practitioner of Artificial Intelligence requires a detailed understanding of how Artificial Intelligence and Machine Learning algorithms work beyond just the mathematics of these algorithms. The mathematics of Machine Learning algorithms are not extraordinarily complex. Solving engineering-related problems using Artificial Intelligence and Machine Learning goes way beyond only mathematics and statistics. It requires an understanding of neurobiology. It requires knowledge of the philosophy of this technology and its differences with our traditional approaches to engineering problem-solving.

As mentioned in an article from Stanford University [11]; "Artificial Intelligence (AI) is the field devoted to building artificial animals ... and, for many, artificial persons ... Such goals immediately ensure that AI is a discipline of considerable interest to many philosophers, and this has been confirmed (e.g.) by the energetic attempt, on the part of numerous philosophers, to show that these goals are in fact un/attainable. On the constructive side, many of the core formalisms and techniques used in AI come out of, and are indeed still much used and refined in, philosophy".

To cut the story short, using Artificial Intelligence experts and/or statisticians that have minimal understanding of engineering domain expertise to solve engineering-related problems usually generates very poor results and, at best, sometime will create highly mediocre outcomes. This has caused many managements of industrial companies to blame the Artificial Intelligence and Machine Learning technology rather than realizing that it was the incorrect decision of using the wrong individuals to guide their company in this area or using the wrong service companies and vendors to solve their problems when they do an incredible amount of marketing of their knowledge and characteristics.

The petroleum industry is one of the top industries that has been making these mistakes for years. Management of the petroleum industry must be correctly guided on how to use Artificial Intelligence and Machine Learning as a positive and fantastic technology for problem-solving and decision-making in their company.

Second Problem: **The involvement of traditional engineers who do not believe Artificial Intelligence is a science and technology and claim that Artificial Intelligence is a hoax (same as Covid-19) and should not be used to solve engineering-related problems.**

Traditionalist engineers refer to those who *believe* the only way to solve an engineering-related problem is through the development of mathematical formulations that can describe the specific physical phenomenon they are trying to solve. They *believe* that what they have learned, experienced, and been dealing with for a long time is the *only* way to solve engineering-related problems. To many of them, this specific, traditional approach to engineering problem-solving is more of a religion than science. They keep saying that "Artificial Intelligence and Machine Learning is a hoax", or "I tried it, but it never works". What is mentioned here is not just an idea, it is an experience. It is about many traditional scientists and engineers (most of them faculty members) who have mentioned such things to me in many years.

Until about a decade ago, these types of engineers faced the toughest problems with the application of Artificial Intelligence and Machine Learning in engineering-related problem-solving. Some of them still exist and do all they can to annoy those engineers and scientists who are interested in learning and working with this innovative technology. Fortunately, given the positive behavior of the new generation of engineers and scientists, the current traditionalist engineers are no longer a problem. Many of them still try to use as much politics as they can to make sure that the application of Artificial Intelligence and Machine Learning in engineering-related problem-solving can be avoided. As time goes on, the number and the problems associated with these types of individuals keep going down.

Third Problem: **The engineers who have a minimal and superficial understanding of Artificial Intelligence are mostly interested in it from a business and marketing point of view rather than science and technology.**

Today, the most important problem related to the application of Artificial Intelligence and Machine Learning in engineering-related problem-solving has to do with some engineering domain experts. One of the main problems associated with the application of Artificial Intelligence and Machine Learning by some engineers is when they try to achieve solutions without following the main characteristics of Artificial Intelligence and Machine Learning. From their point of view, if some mathematics that incorporates machine learning algorithms is used to achieve their solution, then they feel comfortable to call it an AI-based approach. What they miss is correctly answering this question: "Why would you want to use Machine Learning Algorithm to achieve a solution?"

One answer is: "Because people (management of my company or my client) are interested in Artificial Intelligence and Machine Learning, therefore, I can even use Machine Learning algorithm to perform "Linear Regression". Well, this is true. It allows you to exercise Machine Learning algorithm to better learn how to use it. However, this approach has only an "academic" value that allows you to learn, but it does not have a "realistic" value because it does not follow the main reason "WHY Machine Learning algorithm must be used".

Another answer is: "I tried to use actual data (actual measurements), but it did not work. Maybe I did not have enough data, or maybe the actual data was too uncertain and had too much noise, and maybe it was too complex to achieve a solution. Therefore, I decided to generate data from the equations and combine it with the real data, which significantly increased the possibility of success for pattern recognition".

Well, it makes sense because data that is generated by mathematical equations is based on existing correlations, and therefore it guarantees that patterns from this data can be recognized by Machine Learning algorithms. Again, this has an "academic" value that allows you to learn how Artificial Intelligence and Machine Learning algorithms work, but it does not have a "realistic" value because it does not follow the main reason "WHY Machine Learning algorithm must be used".

Let us be clear: trying multiple approaches to achieve reasonable solutions is the right thing to do. However, it is important to be able to correctly judge the essence of the approach that is being used, and it is extremely important to pay attention to the scientific reasoning of "WHY Machine Learning algorithm must be used". Unfortunately, in many cases, the judgment is based on marketing and business success rather than scientific correctness. Under such circumstances, the lack of realistic and scientific success in using Artificial Intelligence and Machine Learning in engineering-related problem-solving is clarified.

It seems that those engineering domain experts who end up doing what was explained above and are also being called "Hybrid Models" have a superficial and limited understanding of Artificial Intelligence and Machine Learning. In general, this specific problem associated with engineering domain experts can be divided into two categories:

1. Those who try to combine data generated by mathematical equations with actual measurements when trying to solve engineering-related problems using Artificial Intelligence and Machine Learning. This is called "Hybrid Model".
2. Those who try to use specific machine learning algorithms that have been developed to solve non-engineering-related problems to solve engineering-related problem. Here are two examples: (a) they use Convolutional Neural Network (CNN) that was developed to solve non-engineering-related problems such as "image recognition", and (b) they use Long-Short-Term Memory (LSTM) that was developed to solve non-engineering-related problems such as "caption generation" for images.

REQUIREMENTS OF ARTIFICIAL INTELLIGENCE FOR SCIENCE AND ENGINEERING

As it has already been mentioned, the major requirements of Artificial Intelligence of science and engineering are two areas of expertise: (1) Expertise in the specific science and engineering topic that Artificial Intelligence is being applied to and (2) Expertise in Artificial Intelligence and Machine Learning. If anyone is missing or is not truly an expert on one of the above two required areas of expertise, then they will not be able to realistically and correctly use Artificial Intelligence to enhance the solutions to science- and engineering-related problems. Unfortunately, this (unrealistic and incorrect application of Artificial Intelligence in science and engineering) has been happening in science and engineering for the past decade throughout the world.

Many things that so far and currently are being mentioned as Artificial Intelligence and/or Machine Learning application in modeling and solving science- and engineering-related problems have included the use of traditional statistics or the same traditional science and engineering approaches through the use of some Machine Learning algorithms. Such approaches are always called "AI-based", which is completely incorrect. The reason that such approaches are not correct is because such individuals and companies have very little understanding of Artificial Intelligence but use this term mainly from a marketing perspective. However, given the importance of the AI revolution in our world, science and engineering applications of Artificial Intelligence will enhance, hopefully soon.

In the past decade, when scientists and engineers have become interested in Artificial Intelligence due to its importance in every aspect of our lives, they started to interact with Artificial Intelligence and use it to solve science and engineering problems. There are five major problems associated with how scientists and engineers in the past decade have been using Artificial Intelligence and Machine Learning: (1) Major Concentration on Mathematical Equations of the Machine Learning Algorithms, (2) Inclusion of Mathematical Equation of the Science- and

Engineering-Related Problem in the AI-based Modeling, (3) Using the Machine Learning Algorithms that have been developed and used to solve Artificial General Intelligence-related problems, (4) Using Traditional Statistical approaches since they are also data-driven, and (5) Continuing the Traditional Engineering approaches to problem-solving but using Machine Learning algorithms. Following are short explanations of each of these five items that have been mentioned:

Problem #1: Major Concentration on the Mathematical Equations of Machine Learning Algorithms

Based on the traditional approach of science and engineering problem-solving, scientists and engineers have been concluding that as long as they understand and learn the mathematical equations of the machine learning algorithms that they will be using, then that makes them experts in Artificial Intelligence. Their conclusion is that they will be able to use Artificial Intelligence to correctly solve science- and engineering-related problems (AI-based), since they know exactly how the mathematical equations of the Machine Learning Algorithms are being used. The main reason for such an understanding is that in traditional engineering, it is an absolute requirement to be able to understand and learn the mathematical equations that have already been developed for any physical phenomena.

Unfortunately, this is a misunderstanding of Artificial Intelligence. It must be understood that using Machine Learning Algorithms to solve science- and engineering-related problems and create AI-based solutions is quite different from the traditional approach to modeling physical phenomena. Using Artificial Intelligence to analyze, model, solve, and optimize science- and engineering-related problems is quite different from our traditional approach to doing this, which is through generating mathematical equations. Given the fact that the mathematical equations associated with the activities of the Machine Learning Algorithms are quite simple (relative to many of the mathematical equations that have developed for physical phenomena), overwhelming scientists and engineers think of becoming AI experts in a very short amount of time since they fully understand and can use the Machine Learning Algorithm. What is incorrect in their understanding of becoming experts applicable to Artificial Intelligence is their lack of realization that they must get a full understanding of Evolution, Neuroscience, and Philosophy that are associated with Artificial Intelligence and Machine Learning Algorithms. More details on this topic will be covered in Chapter 4.

Problem #2: Inclusion of Mathematical Equations of the Science- and Engineering-Related Problem in AI-based Modeling

Currently, the overwhelming majority of scientists and engineers that are trying to model specific science and engineering topics using Artificial Intelligence include the mathematical equations that have been developed for those specific science and engineering topics. Since the Machine Learning Algorithms are data-driven, their inclusion of "Physics-based Mathematical Equations" is done through the generation of data using the mathematical equations that have been developed and used to solve such problems in the past. They generate data from those mathematical equations and combine them with the actual and real data that has been collected for AI-based modeling. Such an approach that has been quite common in many scientific and engineering approaches is usually called "Physics-based" and/or "Hybrid Models".

Since all the variables and parameters that are included in the mathematical equations are correlated to each other, generating data using mathematical equations for such variables and parameters to be included in the data-driven modeling is asking Machine Learning Algorithm to follow (and actually use) the mathematical equations that has generated the data to solve the problem. The same approaches have been happening through the traditional approach of solving science- and engineering-related problems. This approach clearly proves the lack of understanding of the science and engineering applications of Artificial Intelligence.

Solving science- and engineering-related problems through Artificial Intelligence is only based on actual and real data that does not include any assumptions, interpretations, or simplifications. This means that the correct application of Artificial Intelligence to solve science- and engineering-related problems must not include any data that is generated through mathematical equations. Expertise in specific science and engineering topics is required to fully understand and use the actual (measured) and real data to *teach* the Machine Learning Algorithms about physical phenomena.

An AI-based approach is used to produce science and engineering solutions through learning from the actual (measured) and real data and performing pattern recognition by parallel, distributed information processing of the actual (measured) and real data. The mistakes that have been quite commonly used by scientists and engineers in the past decade include data generation using mathematical equations to be combined with actual data that has been collected and measured. More details on this topic will be discussed in Chapter 6.

PROBLEM #3: USING THE MACHINE LEARNING ALGORITHMS THAT HAVE BEEN DEVELOPED AND USED TO SOLVE ARTIFICIAL GENERAL INTELLIGENCE-RELATED PROBLEMS

To a large extent, currently and in the past decade, the overwhelming majority of scientists and engineers end up using the machine learning algorithms that have been developed to solve General Intelligence problems using Artificial Intelligence such as CNNs, LSTM, etc.... The fact is that such algorithms that have been developed and used to solve Artificial General Intelligence such as image recognition, voice recognition, language translations, caption generation, face recognition, object recognition, autonomous vehicles, etc. should not be used exactly in the same fashion to solve engineering-related problems.

CNNs that work for image recognition go through two major processes. The first part is convolution of the image metrics, and the second part is using the Artificial Neural Network to train the convolved image metrics, as shown in Figure 3.8. Convolution is the merging of information through the combination (convolving) of

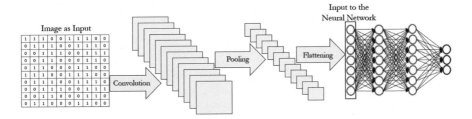

FIGURE 3.8 The process of convolutional neural networks (CNN) that is used for image recognition. CNN is a combination of modifications to the image (convolution using Kernel, Pooling, and Flattening) and neural network.

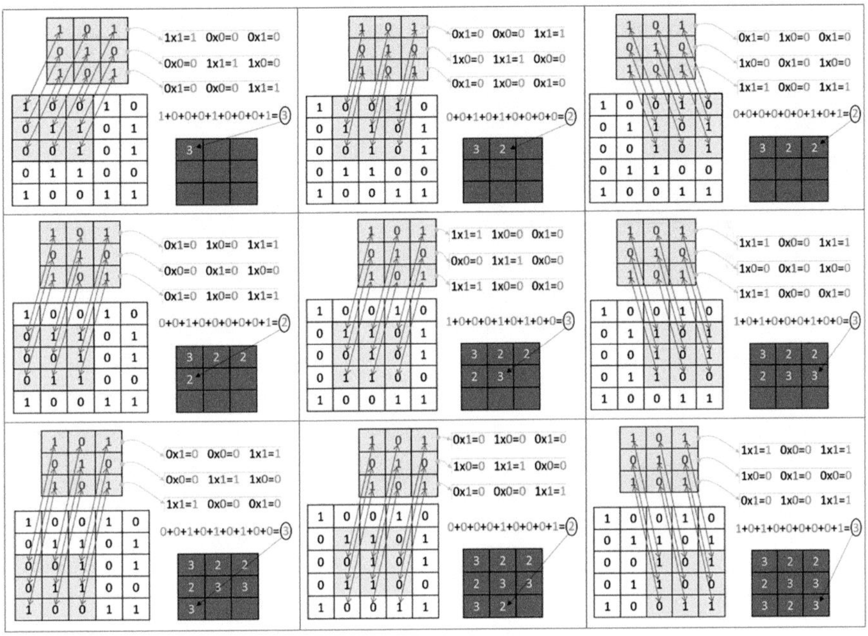

FIGURE 3.9 Convolution layer using Kernel.

mathematical functions to create a new function. As shown in Figures 3.9–3.12, the convolutional process includes several types of filtering processes, such as Kernel, Pooling, and Flattening of the initial metrics of the images that are being used for the training Neural Network.

The objective of convolution is to extract useful features from the images that are being used for the training of the image recognition process. Figure 3.8 shows 10×10 pixels of the image being used, and Figures 3.9 and 3.10 show 5×5 pixels of the images (in 0 and 1) that are being used for the image recognition training.

The example of the Kernel process is shown in Figure 3.9 in a step-by-step process and in Figure 3.10 when the entire process shown in Figure 3.9 has been completed. Figure 3.11 shows the example of Pooling Layers where the maximum value of each

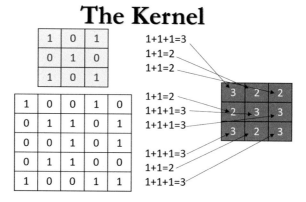

FIGURE 3.10 Convolution is performed on the input image through filtering/Kernel.

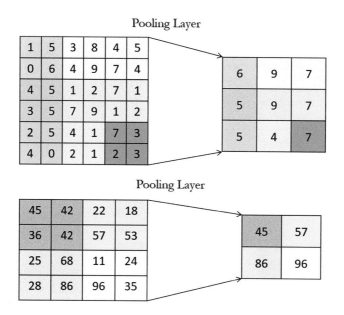

FIGURE 3.11 Pooling layer after the convolution layer on the matrix of the images.

series of the Kernel is used, and finally, Figure 3.12 shows that the Flattening process takes the final calculated matrix into a single input series.

PROBLEM #4: USING TRADITIONAL STATISTICAL APPROACHES SINCE THEY ARE ALSO DATA-DRIVEN

Since the Artificial Intelligence approach is currently called data-driven, a large number of scientists and engineers (specifically in academia) think that Artificial Intelligence approach is the same as Traditional Statistics since it also only uses data to solve problems. Unfortunately, they do not realize that Artificial Intelligence does not use the

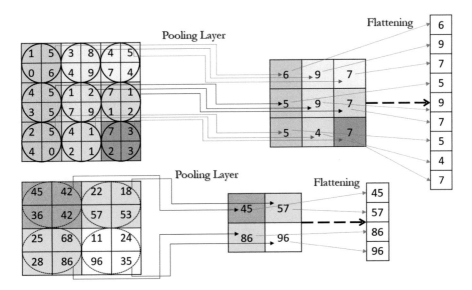

FIGURE 3.12 Flattening is the pooling layer on the matrix of the images.

same techniques of problem-solving that have been used by Traditional Statistics in the past 150 years. They do not realize and may not even be interested in learning the technical and philosophical differences between Traditional Statistics and Artificial Intelligence. More details on this item are discussed in Chapter 6 of this book.

PROBLEM #5: CONTINUING THE TRADITIONAL ENGINEERING APPROACH TO PROBLEM-SOLVING BUT USING MACHINE LEARNING ALGORITHMS

While science and engineering domain expertise is an absolute requirement to solve problems using Artificial Intelligence, a realistic and correct understanding and expertise of Artificial Intelligence are also required. It must be understood that Artificial Intelligence solves engineering-related problems in a quite different fashion than we have been solving science- and engineering-related problems in the past few centuries. It must be learned that when Artificial Intelligence is used to solve science- and engineering-related problems, its objective is to avoid (as much as possible) assumptions, interpretations, and simplifications. This is the fact that Artificial Intelligence separates and enhances the traditional problem-solving of science and engineering. The domain expertise in science and engineering will help to *Teach* these important intelligent items to the machine *Learning* algorithms.

EXPERTISE IN SCIENCE AND ENGINEERING

Since the "problems associated with how Artificial Intelligence and Machine Learning have been used by scientists and engineers in the past decade have already been mentioned", the next item is to clarify what the actual requirements of Artificial Intelligence Application in science and engineering. As mentioned at the start of this chapter and as shown in Figure 3.13, the requirements of Artificial Intelligence for science and engineering applications are: (1) Expertise in the specific science and engineering topic that Artificial Intelligence is being applied to and (2) Expertise in Artificial Intelligence and Machine Learning.

As shown in Figure 3.14, in the past decade, all engineering industries started using Artificial Intelligence experts who did not have any expertise in specific engineering

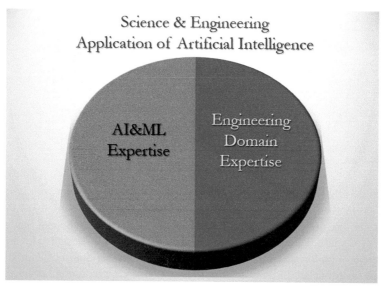

FIGURE 3.13   Characteristics of engineering applications of Artificial Intelligence.

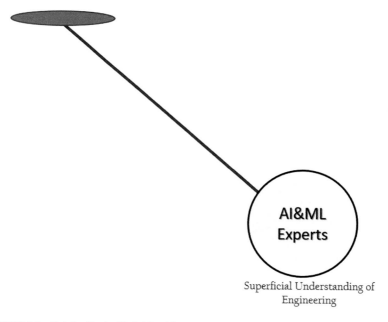

FIGURE 3.14   Originally Artificial Intelligence experts were trying to model and solve artificial engineering intelligence problems and the results were not good at all.

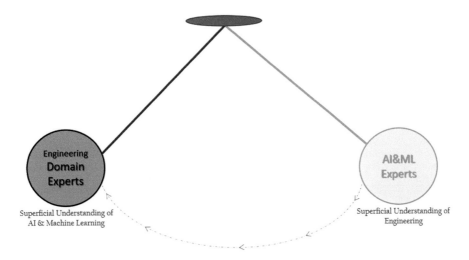

FIGURE 3.15 Later, engineering domain experts were trying to model and solve Artificial Engineering Intelligence problems without having expertise in Artificial Intelligence and the results are still not good at all.

to solve their problems. Then, when the results of the AI-modeling and problem-solving ended up being very poor, the idea of the importance of domain expertise in engineering to solve the problem using Artificial Intelligence became so important. However, in the past several years (and even now and may be in the next few years), the overwhelming engineering domain experts that started using Artificial Intelligence have had their results and approaches just as poor as before. The main reason for such problems, as shown in Figure 3.15, has to do with their superficial understanding of Artificial Intelligence and Machine Learning. As engineers, they think using Artificial Intelligence to solve engineering-related problems is all about knowing the mathematical characteristics of the Machine Learning Algorithms. It is the reason that the overwhelming majority of them call their approaches Machine Learning and not Artificial Intelligence. Both of these cases that have been presented on the right and left sides of Figures 3.14 and 3.15 need to be avoided. The best way to be able to use Artificial Intelligence for science and engineering problem-solving is shown the middle of Figure 3.16. Hopefully, the pendulum swings to the right and left and eventually ends up in the middle.

Expertise in science and engineering does not refer to the inclusion of the mathematical equations that have been developed for the specific physical phenomenon that is being solved using Artificial Intelligence. Rather, the most important expertise in science and engineering is the understanding of the acts and performances of a specific physical phenomenon. The expertise will include intelligence about the worst things that are negative about this specific physical phenomenon and the best things that are possibly positive about this specific physical phenomenon. The actual experience of interacting with such specific physical phenomena is the most important part of the expertise mentioned here.

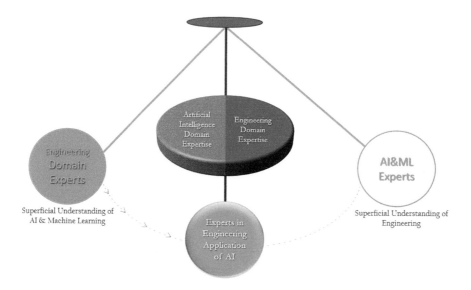

FIGURE 3.16 Pendulum swing of engineering application of Artificial Intelligence.

Such expertise will be used to identify the best actual (measured) data that must be used as input to the machine learning algorithms and to "Generate" new "Features" from the actual data that has been measured and provided to solve this specific physical phenomenon. This is the way that science and engineering expertise are used to *Teach* the reality to machine *Learning* algorithms through *Data*. In Machine Learning, teaching cannot be done through speaking, typing, assuming that the algorithm knows math and physics, etc. The only way to be able to *Teach* the Machines so that they can *Learn* is by using the actual measured *Data* that is provided to solve the physical phenomenon using Artificial Intelligence.

Please note that when it is mentioned "Feature Generation", it does not mean that any mathematical equations should be used to generate new data. This specific type of "Feature Generation" is used to "Teach" the Machine Learning Algorithm about the specific physical phenomenon that is being addressed using the existing actual (measured) data. This means that the actual measured data that is provided for this specific physical phenomenon should be used to generate new series of data based on our science and engineering applications.

What is incredibly important about the science and engineering expertise is the strong knowledge about the problem that is being addressed while having an open mind about the innovative approaches for generating the solutions. Unfortunately, I have experienced some experts in certain science and engineering topics who do not believe it is possible to do anything else other than what they have been doing for decades to solve a specific type of problem. Such individuals seem to have a religious mind about the technology that they have already learned and worked with for several years and do not believe that anything else could possibly be provided to enhance the solutions to the same problems that they have been dealing with.

EXPERTISE IN ARTIFICIAL INTELLIGENCE

The traditional approach to modeling physical phenomena using science and engineering that has been done in the past few centuries is through mathematical equations. This causes a lot of scientists and engineers to think that all they need to become experts in Artificial Intelligence is to know the details of the mathematical equations about machine learning algorithms that are used to generate Artificial Intelligence. The fact is that having good and detailed information about the mathematical equations of machine learning algorithms does not mean that it makes scientists and engineers experts in Artificial intelligence.

It is a well-known fact that a good amount of time (several years) has been required for you to become an expert in science or engineering. Then, why would you think that the same amount of time, studying, researching, and experience is not required for you to become an expert in Artificial Intelligence? I am sure that what has been shown in Figure 3.17 makes absolutely no sense. Actually, the fact is that no one and no company has ever done the type of marketing that I have done, as shown in Figure 3.17. The reason I have made this figure (Figure 3.17) is to demonstrate how unrealistic such marketing can possibly be. And, as I have already mentioned, no one has ever done such marketing. However, the reason Figure 3.17 was made up was to have the same understanding and lack of belief as what is shown in Figure 3.18.

FIGURE 3.17   Does it make sense to anyone that it is possible for you to become a master (an expert) in a major science and engineering (physics, mechanical engineer, petroleum engineer, geologist, chemical engineer, reservoir, engineer, etc....) in only 1 month?

FIGURE 3.18   Does it make sense that becoming a "Data Scientist" can only be done in 1 month?

What is shown in Figure 3.18 is not made-up. It is absolutely not correct, but actually, it is a fact. This has been done by a company (UDEMY) that makes a lot of people think that they need very little time and effort in order to become experts in Artificial Intelligence, which is now is also being mentioned as "Data Science".

Expertise in Artificial Intelligence requires a large amount of research and development of problem-solving skills using this new modern technology. The research and development of problem-solving using Artificial Intelligence will incredibly interest scientists and engineers in evolution, philosophy, neuroscience, and biology, as shown in Figure 3.19.

    a. *Evolution* provides information about the enhancement and improvement of the brain of Home sapiens. Beside the comparison of the brains of all the animals that were developed before us, it even lets us compare the strength of our brains with the brains of other human species' such as Neanderthals and *Homo erectus*. Neanderthals and *Homo erectus* evolved hundreds of thousands of years before us but never left the hunter-gatherer characteristics.

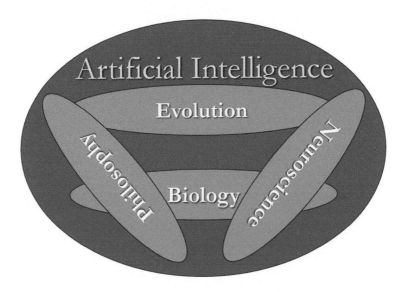

FIGURE 3.19 Fundamental knowledge and understanding of Artificial Intelligence and Machine Learning require interest and expertise in philosophy, neuroscience, mathematics, computer science, statistics, and biology.

> *Homo sapiens* left eastern Africa almost 70,000 years ago, changed their hunter-gatherer characteristics through agricultural revolutions about 12,000 years ago, and started the creation of science and industry about 350 years ago. Without an interest in evolution, it would be very hard and unrealistic to learn about the fantastic characteristics of the human brain.

b. *Philosophy* provides incredible information and understanding about the fundamental nature of knowledge, fact, reality, and experience. As with the traditional approach to science and engineering, Aristotelian philosophy must not be the only approach used in Artificial Intelligence.

c. *Neuroscience* as a fantastic technology focuses on the human brains' activities and behaviors through neurons.

d. *Biology*, which is the combination of two Greek words, "bios", which means "life", and "logos", which means "study", is the science of life and living organisms.

Given the definition of Artificial Intelligence, which is the replication of human intelligence through mimicking the "Human Brain", its expertise will make scientists and engineers to become incredibly interested in Evolution, Philosophy, Neuroscience, and Biology. It is highly important to note that once you become capable of using machine learning algorithms (which are quite simple to do for scientists and engineers), it does not mean in any shape or form that you have become an Artificial Intelligence expert. Unfortunately, this is currently one of the main problems for scientists and engineers: when they start using machine learning algorithms, it makes them feel there is no need to learn more about Artificial Intelligence.

NOTE

1 I was not allowed to copy the figure of these online "News Releases" in this book. However, if you do not believe that this is true, then you can send me an email and ask for these "online News Releases" that I can send them to you since I took pictures of them and it seems they have been removed from the Internet.

4 Modeling Physics Using Artificial Intelligence

Modeling physics using Artificial Intelligence involves facts, reality, and actual measurements while avoiding assumptions, interpretations, and simplifications. In this chapter, we start with a short repetition of the last chapter and then concentrate on the topic of this chapter. It is quite interesting to note and discuss the topic of differences between developing models of "Physical Phenomena" and developing models of "General Phenomena" through Artificial Intelligence. Furthermore, it is important to note: "If there are differences between solving physics-based problems and general problems using Artificial Intelligence"? To be able to correctly answer these questions, it would be interesting to first respond to another very simple question. The response to this very simple question would be quite easy and would be helpful to answer the first two questions (mentioned above in this paragraph) about Artificial Intelligence.

Here is a simple question that anyone can answer quickly.

QUESTION 1

Using "Human Intelligence", are there any differences between modeling and/or solving "Physical Phenomena" versus modeling and solving "General Phenomena"? A reasonable example of this question would be: using "Human Intelligence", are there any differences between "Modeling CO_2 Injection into Saline Aquifer for the positive impact of Climate Change" versus "Identification of Images such as the difference between a Cat and a Dog"?

ANSWER 1

Yes, indeed. There are differences between modeling and/or solving physical phenomena and general phenomena using "Human Intelligence". To model or solve physical phenomena, humans need to go to school and university for several years and become experts in the specific type of physical phenomena that they need to solve. "Human General Intelligence" is good enough to solve "General Phenomena".

The above answer about "Human Intelligence" is very straight and quite simple to agree with. Since "Artificial Intelligence" is a process to use machine learning algorithms to simulate "Human Intelligence", it must go through the same process as "Human Intelligence" to model and/or solve physical phenomena. When this is mentioned, many individuals will then ask:

DOI: 10.1201/9781003369356-5

QUESTION 2

How can we make machines (computers – machine learning algorithms) to go to school and university for several years and become experts in the specific type of physical phenomena that is needed to be solved through "Artificial Intelligence"?

That is a good and correct question. Here is the answer to this question:

ANSWER 2

First and foremost, machines (computers – machine learning algorithms) do not need (and of course it's impossible) to go to school and university to be experts in physical phenomena. But there are other ways to "Teaching" Machine "Learning" algorithms. Since "Learning" of physical phenomena requires "Teaching", the teacher of physical phenomena to machine learning algorithm must be an expert in the specific physical phenomena topic. The next important item would be about the techniques that the physical phenomenon expert must use to "Teaching" Machine "Learning" algorithms. It is important to note that this "Teaching" to the Machine "Learning" algorithms would be all about dealing with the available facts, reality, and actual measured data.

Therefore, it is absolutely important to note that when modeling and/or solving physical phenomena using "Artificial Intelligence", not only expertise in "Artificial Intelligence" must be included, but also expertise in the specific type of physical phenomena that is being addressed must be included by those individuals that are going to use "Artificial Intelligence". As shown in Figure 4.1, both "Human Intelligence" and "Artificial Intelligence" require having "General Intelligence" to solve and model "General Phenomena" and "Domain Expertise" to solve and model "Physical Phenomena".

There are two important topics that must be covered in this chapter to explain how "Artificial Intelligence" must be used to model and/or solve physical phenomena. These two topics are:

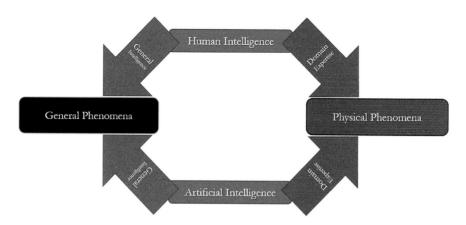

FIGURE 4.1   Requirements of general intelligence and domain expertise in both human intelligence and artificial intelligence.

A. The differences between traditional modeling of physical phenomena and AI-based modeling of physical phenomena, and

B. How domain expertise in specific physical phenomena must be used through Artificial Intelligence to model and/or solve those physical phenomena.

As mentioned above, the first topic is the uniqueness and differences between Artificial Intelligence and the techniques that have been used in the past few centuries to model and/or solve physical phenomena, and the second topic identifies how it must be done.

ARTIFICIAL INTELLIGENCE VERSUS TRADITIONAL MODELING IN PHYSICS

What we have been learning in schools and universities to model and/or solve physical phenomena can be called "Traditional Modeling and Solving Physical Phenomena". This "Traditional Technology" uses mathematical equations to model physical phenomena. The major differences between Traditional Technology and Artificial Intelligence (brand new technology) for modeling and/or solving physical phenomena have to do with using and/or avoiding mathematical equations for modeling the physical phenomena. While mathematical equations become the model of physical phenomena through traditional science and engineering technology, Artificial Intelligence does not model physical phenomena through mathematical equations. The main reason that Artificial Intelligence does not use mathematical equations to model and/or solve physical phenomena is the avoidance of any assumptions, interpretations, and simplifications. Artificial Intelligence models and/or solves physical phenomena that are always purely based on reality, facts, and actual measurements.

In general, the "Traditional Modeling and Solving Physical Phenomena" that has been used in the past several centuries includes the identification of variables and parameters that play a role in the specific physical phenomenon. Then the identified variables and parameters are used for the development of a mathematical equation. When such mathematical equations are not too complex, then exact solutions can be generated through analytical solutions. However, when certain physical phenomena are modeled through highly complex mathematical equations, analytical solutions cannot be used. In such cases, numerical solutions have been developed to solve the complex mathematical equations.

As it was covered in the previous chapters, Artificial Intelligence is defined as the simulation of "Human Intelligence" and since "Human Intelligence" is generated through the "Human Brain", mimicking "Human Brain" activities would be the most important part of Artificial Intelligence. Based on the simple definition of traditional modeling of physical phenomena, which is the use of mathematical equations for modeling physics, the question that can be asked is: Does "Human Brain" use mathematical equations to solve or interact with any physical phenomena? The answer to this question is completely easy and obvious. To cover an example for this question, let's use what is shown in Figure 4.2.

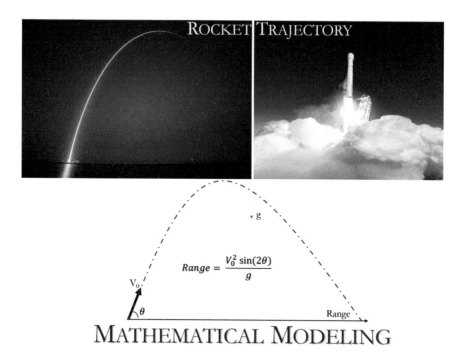

FIGURE 4.2   Identifying the final dropping location of the rocket trajectory by physics and mathematics.

This figure shows a physical phenomenon (rocket trajectory) that has been modeled using mathematical equations. The objective of the rocket trajectory is to reach a specific location that needs to be hit and destroyed. To identify where the rocket that is being thrown will be coming down, a physical model through mathematical equations is used, as shown in Figure 4.2. To reach the location where the rocket will be coming down, it is a function of gravity (g), initial velocity (V_0), and the initial angle (θ) for shooting the rocket. However, looking at Figure 4.3, the same action is taking place between humans.

One human throws an item at another human. In a football game, the quarterback will throw the ball to be caught by the receiver. The question is: Does the brain of a human (the receiver) that must catch the item (football) also generate a mathematical equation to identify where the item (football) will come down? If that is the case, then the human brain after generating the mathematical equation, must solve it. Based on the solution to the mathematical equation that the human brain achieves, should the brain of the receiver guide his feet and arms to act and catch the football?

It seems that the answer to this question should be obvious. The fact is that the human brain does not develop mathematical equations to solve a physical phenomenon. Since that is the case (not developing and solving mathematical equations), then what does the human brain do to solve this physical phenomenon? How will the receiver on the U.S. football team be able to identify where the football comes down so that he can use his feet and arms to catch the football?

FIGURE 4.3   Identifying the final dropping location of the ball trajectory by the human brain.

Human eyes look at the football and even the way the Quarterback activated to throw it to him, and then using such "Data" identifies which direction to go and how far to be able to catch the football. In other words, in many cases, "Natural Intelligence" through the "Human Brain" tries to solve a problem that is a physical phenomenon. "Natural Intelligence" of the "Human Brain" uses the actual data that it observes in order to solve the physical problem. "Natural Intelligence" of the "Human Brain" does not develop a mathematical equation and then try to solve it. If the activity of the "Human Brain" through "Natural Intelligence" in this fashion makes sense, then shouldn't "Artificial Intelligence" activate in the same fashion to solve similar problems? The answer is: **ABSOLUTEY YES**. Artificial Intelligence must simulate Natural Intelligence of the human brain to model and solve physical problems through actual data. Another example of such a situation (solving a physical phenomenon) is shown in Figure 4.4.

Artificial General Intelligence technologies have been used for face, object, and voice recognition, as well as autonomous vehicles, human conversation, and large language model (LLM). It is quite obvious that the "Human Brain" does not need any physics-based domain expertise (through going to school and university for learning) to be able to solve non-physics-based problems. Identifying faces and voices or different objectives, such as cars, buses, trucks, bikes, pedestrians, … in the streets or being able to learn how to drive a car and have language conversation, does not require anything other than General Intelligence in the human brain. While it is identified and mentioned that many of the current fantastic AI-based technologies that have been developed are based on General Intelligence, it is also very important

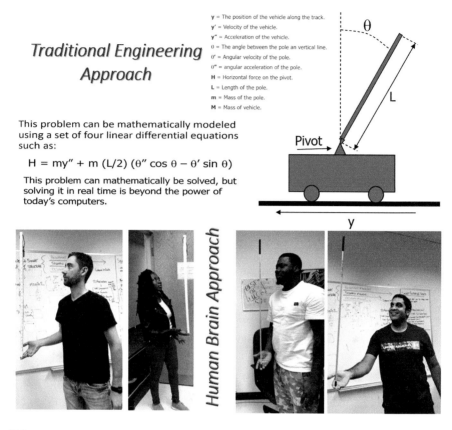

Traditional Engineering Approach

y = The position of the vehicle along the track.
y' = Velocity of the vehicle.
y" = Acceleration of the vehicle.
θ = The angle between the pole an vertical line.
θ' = Angular velocity of the pole.
θ" = angular acceleration of the pole.
H = Horizontal force on the pivot.
L = Length of the pole.
m = Mass of the pole.
M = Mass of vehicle.

This problem can be mathematically modeled using a set of four linear differential equations such as:

$$H = my" + m (L/2) (\theta" \cos \theta - \theta' \sin \theta)$$

This problem can mathematically be solved, but solving it in real time is beyond the power of today's computers.

Human Brain Approach

FIGURE 4.4   Balancing a stick by physics and mathematics versus the human brain.

to note that in this book it is not claimed that these Artificial General Intelligence technologies are simple and easy to develop.

It is quite important to note that developing Artificial General Intelligence technologies is not easy or simple since they are simulating "Human General Intelligence", which does not require science and engineering expertise. The fact is that the overwhelming majority of these Artificial General Intelligence technologies were quite complex during their development, and their expansion requires incredible expertise in Artificial Intelligence. Nevertheless, another fact is that while these AI-based experts are creating Artificial General Intelligence technologies, they are not technologies that would be able to generate models for specific physical phenomena that, in the past centuries, have been done through mathematical equations. These mathematical equations of the physical phenomena are then solved through analytical solutions, or numerical solutions, based on the complexities of the mathematical equations that have modeled these physical phenomena. It is obvious that a lot of important issues that have taken place in *Homo sapiens* only needs "Human General Intelligence". Therefore, it is very important to be an expert in "Artificial Intelligence" to be able to develop "Artificial General Intelligence" models.

Recent technologies such as "ChatGPT", "GPT-4", "Google Bard", … are part of "Artificial General Intelligence" and it is extremely important to note the importance of "Artificial Intelligence". However, it does not mean that these "Artificial General Intelligence" technologies are able to solve science- and engineering-related problems through modeling physical phenomena.

HOW PHYSICAL DOMAIN EXPERTISE MUST BE USED FOR AI-BASED MODELING OF PHYSICS

One of the major issues that will be included in this chapter is the way physical domain expertise must be used to develop physics-based Artificial Intelligence models. First and foremost, it must be mentioned that (A) physics-based Artificial Intelligence models will be quite different from traditional modeling of physical phenomena by engineers and scientists, (B) physics-based Artificial Intelligence models can be developed much quicker during development and then far faster to be run compared to the traditional modeling of physical phenomena by engineers and scientists, and (C) physics-based Artificial Intelligence models will be providing much better and far more realistic results than traditional modeling of physical phenomena by engineers and scientists. First, physical domain expertise will be discussed in the context of these three items that have been mentioned. Once the difference between traditional physical modeling and AI-based modeling of physical phenomena is covered, the way domain expertise must be used to develop science- and engineering-related models using Artificial Intelligence will be covered.

DIFFERENCE OF PHYSICS-BASED AI MODELS WITH TRADITIONAL PHYSICS-BASED MODELING

It is a well-known fact that traditional physics-based models are done using mathematical equations. It has already been mentioned that scientists and engineers identify the variables and parameters that are involved in a physical phenomenon and then use them inside a mathematical equation to model it. When Artificial Intelligence is used to develop a physics-based model, it is all based on the actual data. This is the main reason that it has also been called data-driven, data-science, and data-analytics. However, what is most important about this item is that currently it has not been correctly used by many scientists and engineers. The most important item in using Artificial Intelligence to develop physics-based models is ONLY using "Actual Data Measurements" for AI-based model development rather than the inclusion of data that has been generated using mathematical equations. More details on this item will be covered in Chapter 6 of this book.

One of the major problems that is currently being addressed by a large number of scientists and engineers is the combination of data that has been generated through mathematical equations with some of the "Actual Data Measurements" in order to develop a physics-based model using Artificial Intelligence. The question may be, "Why large number of scientists and engineers add mathematically generated data to the Actual Data Measurements in order to build AI-based models of physical phenomena?"

It would be interesting to ask them why they do that. In my opinion, the interesting question that must be first asked: "Were you successful in building an AI-based model of physical phenomena only using Actual Data Measurements?", if their answer is "yes", then why would they add the data that is generated from mathematical equations to build the AI-based model? The fact is that the reason they use the data that is generated from mathematical equations and combine it with the Actual Data Measurements for building the AI-based model of the physical phenomena is because they have never been able to develop reasonable models and solutions for the physical phenomena using Artificial Intelligence through Actual Measured Data. The definition of "Hybrid Modeling" is the lack of success in using the real characteristics of Artificial Intelligence to develop models and solutions to physical phenomena or a complete lack of understanding of what "Artificial Intelligence" really means.

QUICK DEVELOPMENT OF PHYSICS-BASED AI MODELS COMPARED TO TRADITIONAL PHYSICS-BASED MODELING

It is a quite well-known issue that the complexity of the physical phenomena will result in the complexity of mathematical equations that will be used to model them. Furthermore, it is also a well-known fact that highly complex mathematical equations cannot be solved simply and quickly using analytical solutions. This is the reason why numerical simulation was created in 1946 by John von Neumann and Stanislaw Ulam. To a very large extent, numerical simulation slightly modifies (simplifies) the characteristics of the physical phenomena in order to be able to generate the solution of the mathematical equation.

This process is done by dividing the characteristics of the physical phenomena into many spaces (millions of cells) and into many time steps. Given the fact that a very large number of cells and time steps must be used to generate the solution of a numerical simulation, and then after the completion of the solution to be used to generate a full understanding of what needs to be done to enhance the physical phenomenon's behavior, it always takes a long amount of time, even when a large number of computers are used to perform the solutions.

As you can currently see and test the characteristics of the Artificial General Intelligence models (examples such as Image Recognition, or ChatGPT) you know that the results generated by the AI-based model are incredibly faster than what is done by humans. The same is true about the Science and Engineering Applications of Artificial Intelligence. Physics-based models generated by Artificial Intelligence even for highly complex physical phenomena that would take tens of hours to be done traditionally, will only take a few seconds or a couple of minutes to be solved by Artificial Intelligence.

PHYSICS-BASED AI MODELS PROVIDE MUCH BETTER RESULTS THAN TRADITIONAL PHYSICS-BASED MODELING

As it was mentioned before, using Artificial Intelligence to model physical phenomena requires using actual measured data and not any mathematical equations. It is a fact that when mathematical equations are used to model physical phenomena, assumptions and interpretations become part of the solutions. Furthermore, not every single variable or parameter that is observed in physical phenomena would definitely become part of the mathematical equation. On the other hand, when Artificial intelligence is used to develop a model for physical phenomena, every single actual measurement that has been generated will be included in the AI-based model. While only actual measurements are being used and no mathematical equations are being used, no assumptions, interpretations, or simplifications will become part of the AI-based model of physical phenomena.

This fact always results in a much better and more realistic model and solution for physical phenomena. When the AI-based model of the physical phenomena is used to match the history of the physical phenomena and then used to optimize them, it provides a lot more realistic results than any ideas that have been included in the models. Given that everything is about fact and reality and avoids any assumptions and interpretations, the results are always far better (as well as much, much faster) than traditional mathematical-based models.

Why Physics-Based Domain Expertise Must Be Used for Artificial Intelligence

It is important to note that using Artificial Intelligence to model scientific and engineering physical phenomena only requires data. Artificial Intelligence will learn all the details of physical phenomena through the data that is provided to the machine learning algorithm. The most important issue that must be noted is the use of ACTUAL data on physical phenomena, not any data that can be generated by anyone, even domain experts. Since traditional modeling of physical phenomena has been performed through mathematical equations, the new Artificial Intelligence modeling of physical phenomena will not (must not) use mathematical equations to model them.

Using physical domain expertise to generate Artificial Intelligence models of physical phenomena has several important requirements: (A) avoiding the traditional physics-based modeling that includes mathematical equations, (B) teaching and communicating the major information of the physical phenomena to the machine learning algorithm that is being used to generate Artificial Intelligence model through actual data. Using the physical domain expertise, the scientists and engineers must provide explanations of the important parts and realistic characteristics of the physical phenomena to the machine learning algorithms through the actual measured data that is used for the physics-based Artificial Intelligence modeling.

Domain experts of any specific engineering topic must realize that "Learning" by the machine about the specific engineering topic requires "Teaching". First, "Teaching" can only be provided by specific domain experts, and second "Teaching" must be performed only through actual measured data, not through any data that is generated using any mathematical equations. Since the "First" item ("Teaching" by the specific domain experts) makes sense, the "Second" item ("Teaching" by actual measured data) usually presents some issues to the domain experts. What they need to do is interact with the actual measured data to provide potential information to the machine learning algorithms. Since we have addressed this issue in developing

Reservoir Simulation and Modeling in Petroleum Engineering, it makes sense to provide such example from this engineering technology.

Using Artificial Intelligence to perform Reservoir Simulation and Modeling in Petroleum Engineering, only actual field measurements are used. In petroleum engineering, for the production of oil and gas from reservoirs that are tens of thousands of feet under the ground, a large number of wells (almost a hundred) are drilled from the surface to the reservoir. The actual field measurement data that is used in AI-based Reservoir Modeling includes "operational conditions of each well on the surface", "detail characteristics of how each well has been drilled and perforated", "all the actual measurements of reservoir characteristics through well logs, cores, and seismic data", and "all oil, gas, and water productions".

Such actual field measurement data is generated and used for every single well in a field. Traditionally, reservoir simulation and modeling in petroleum engineering were developed using complex mathematical equations almost 6–7 decades ago and have been solved through numerical simulation. Since we are using Artificial Intelligence instead of traditional modeling or traditional statistics, it will not be successful to only provide such data (actual measurements) for all the wells to the machine learning algorithm. Since reservoir engineering experts know that each well in the field (reservoir) is interacting with several offset wells (around it called offset wells), the domain expert must "Teach" this important issue by using the actual field measurements (not through any data generated from mathematical equations) with machine learning algorithms that are used for AI-based Reservoir Simulation and Modeling. Without doing this "Teaching" (using actual field measurement data), you will not be teaching the machine learning algorithm that the production from each well is not only a function of all the measured data that has been generated for each well, but some actual measured data from each offset well must also be included in every single well so that its production can be generated by Artificial Intelligence in a correct fashion. This is an explanation of how engineering and scientific domain expertise must be used to help Artificial Intelligence.

In the many times, that I have explained the use of "Actual Measured Data" in Artificial Intelligence for modeling physical phenomena, many top scientists and engineers have asked very interesting and correct questions. Their question has been: "Do you realize that all the actual measured data are never 100% accurate? Do you realize that the overwhelming majority of the actual measured data usually have some issues such as uncertainty or noise?" I always reply to them, "Yes, indeed". "You are absolutely correct". In order to answer that question and explain to them that while the "actual measured data are never 100% accurate", they are still the most important information that must be provided to Artificial Intelligence and show them how such an issue does not mean that Artificial Intelligence is unable to use it in a fashion to solve physics problems correctly. To show and explain this issue, I provide them with a paragraph that has been shown in Figure 4.5 and is located in the next page in this chapter.

I am sure that you have already seen the paragraph that is shown in Figure 4.5. I am also sure that you were able to read that paragraph and fully understand what the topic was about. Most probably somewhere in the middle of reading this paragraph, you have noted that something is quite wrong with it, and you then noted that a large number of words that you were reading and understanding were not written correctly. The correct version of the same paragraph is shown in Figure 4.6. All the red-colored words in Figure 4.6 were the words that were incorrectly written in the paragraph in

> Aoccdrnig to a rscheearch at an Elingsh uinervtisy, it deosn't mttaer in waht oredr the ltteers in a wrod are, the olny iprmoetnt tihng is taht frist and lsat ltteer is at the rghit pclae. The rselut can be a toatl mses and you can sitll raed it wouthit porbelm. Tihs is bcuseae we do not raed ervey lteter by it slef but the wrod as a wlohe.

FIGURE 4.5   Please read this paragraph. As you can see, this paragraph includes a total of 67 words. Out of the 67 words, 4 are single letter, 16 are two letters, 13 are three letters, and 34 are more than three letters. Most likely, you will be able to read it completely and then notice that it was not written correctly.

> According to a research at an English university, it doesn't matter in what order the letters in a word are, the olny important thing is that first and last letters is at the right place. The result can be a total mess and you can still read it without problem. This is because we do not read every letter by itself but the word as a whole.

FIGURE 4.6   This is the same paragraph that was shown in Figure 4.5. This is a correctly typed paragraph. You can note that 97% of the words that are more than three letters have been incorrectly typed, but you (most probably) had very few issues correctly reading and understanding this paragraph.

Figure 4.5. Nevertheless, the human brain, which could have read these words, had very little trouble understanding the entire paragraph.

When engineers mention items such as "Actual measured data are never 100% accurate ..." I usually provide them with the paragraph that is shown in Figure 4.5, and then I ask them if they understand what the paragraph is mentioning. The fact is that usually 99.999% of all people are able to read the paragraph shown in Figure 4.5 and also identify that a large number of words in this paragraph are not correctly printed. This means that the human brain is able to understand and deal correctly with "actual data that are never 100% accurate ..." and since Artificial Intelligence simulates natural intelligence through mimicking the human brain, it is also able to handle some of the inaccuracies associated with actual data.

When Artificial Intelligence is used to model and/or solve physical phenomena, its actual data should be presented to the machine learning algorithm in a fashion that represents physics. Those individuals and/or companies that have come to the following three conclusions can be trusted to be able to enhance their objectives using Artificial Intelligence:

a. Domain expertise is a requirement for the engineering application of Artificial Intelligence and Machine Learning,
b. The development of comprehensive AI-based models for engineering-related problems requires *ONLY* actual data (actual measured data), and

    c. There is no need to include any mathematical formulation of physics in this process or add data that is generated through mathematical equations.

Those individuals and/or companies that are including the following items in the work that they call "Data-Driven", or, "Machine Learning", or even "Artificial Intelligence" (lots of them usually do not directly call it Artificial Intelligence), cannot be trusted to be able to enhance their objectives using Artificial Intelligence:

    a. They do not understand or have not come to the conclusion that the engineering application of Artificial Intelligence is different from the non-engineering (general) application of this technology.
    b. They have failed to develop comprehensive and explainable "AI-based Models" for engineering-related problems using *ONLY* actual data (actual measured data) unless additional data is generated through mathematical equations.

Data should be presented to the Machine Learning algorithm in a fashion that represents the physical phenomena that are supposed to be modeled through Artificial Intelligence.

5 Artificial Intelligence versus Traditional Statistics

Since both Artificial Intelligence and Traditional Statistics use data to generate their solutions, it does not mean that they are doing the same thing. The fact is that what they both do is quite different from each other. The current version of Artificial Intelligence, which was developed less than 30 years ago, has generated the types of solutions that have made it the most interesting technology. However, Traditional Statistics, which has been using data to develop models and generate solutions, was developed more than 350 years ago. Therefore, using Traditional Statistics approaches as "Machine Learning" to perform what is called Artificial Intelligence has nothing to do with reality, as will be explained in this chapter. Nevertheless, many current scientists and engineers that have worked with Traditional Statistics in the past several years and now have a good experience with it continue to use, and instead of calling it Traditional Statistics, they now call it Artificial Intelligence and Machine Learning since they are using data.

Some people may wonder why this book includes a chapter about the differences between Artificial Intelligence and Traditional Statistics. The reason has to do with the fact that there are many engineers in academia and industry who believe these two technologies are very much the same because both deal with "Data". When they refer to "Data Analytics", they mostly refer to linear, non-linear, or multivariate regression and try to treat them as a new technology. Many times, people who either intentionally or unintentionally make such mistakes also try to refer to these traditional and old technologies as something that is quite new in their industry.

What they seem to forget is that the use of Traditional Statistical approaches in engineering-related industries dates back to the early 1960s. A good example is the use of Traditional Statistical in petroleum industry. Arp's Decline Curve Analysis in petroleum engineering is a good example of statistical regression. Capacitance-Resistance Modeling (CRM) is another example of using traditional statistical analysis in petroleum engineering to analyze production data. Similar techniques have also been developed in many other engineering technologies. The initial use of geo-statistics in the mining industry, which was later incorporated into numerical reservoir (geological) modeling, dates back to the early 1950s. Therefore, as far as the application of Traditional Statistics in the science and engineering industries is concerned, there is nothing new about this technology.

As was mentioned, Traditional Statistics have been around for more than three centuries. The term "statistic" was originated and invented in Germany in 1749. If the connection of data with probability theory (randomness) is considered, then its

DOI: 10.1201/9781003369356-6

history may even go as far back as the 16th century. Nevertheless, the point is that, unlike Artificial Intelligence, Traditional Statistics are not a new technology. In order to develop a better understanding of the fundamental differences between these two technologies, one should start with the seminal paper written by Leo Breiman, a well-known Statistics Professor from the University of Berkeley [12], followed by the book written by Nisbet, Elder, and Miner (*Handbook of Statistical Analysis & Data Mining Applications*) [13].

As mentioned by the Statistic Professor of the University of Berkeley (Dr. Leo Breiman) in his article [12], "The focus in the statistical community on data models has led to irrelevant theory and questionable scientific conclusions … a predefined pattern (i.e., the parametric model) uses a model to characterize a pattern in the data calls data model". "This enterprise (*Traditional Statistics*) has at its heart the belief that a statistician, by imagination and by looking at the data, can invent a reasonably good parametric class of models for a complex mechanism devised by nature, … truisms have often been ignored in the enthusiasm for fitting data models". "When a model is fit to data to draw quantitative conclusions, the conclusions are about the model's mechanism and not about nature's mechanism". "The belief in the infallibility of the data models was almost religious".

Traditional Statistics has an Aristotelian approach (two-valued logic) to problem-solving and a deductive approach to the truth, while the philosophy behind Artificial Intelligence includes multi-valued logic and an inductive approach to the truth. While Traditional Statistics tests hypotheses through parametric models and compares them to the standard metrics of the models, Artificial Intelligence builds models using the data instead of starting with an existing model and testing the data to see if it fits the predetermined models. Artificial Intelligence is defined as mimicking nature, life, and the human brain to solve complex and dynamic problems. Artificial neural networks mimic neurons in the human brain; fuzzy set theory mimics multi-valued human logic and reasoning; and genetic algorithms mimic the Darwinian evolution theory. Machine learning is defined as a set of open computer algorithms to learn from data instead of explicitly programming the computer and telling it exactly what to do (as engineers do in numerical simulation). The human brain does not use Aristotelian two-valued logic (e.g., yes/no, black/white, true/false, 0/1) to solve problems and make decisions. Probability theory addresses uncertainties that are associated with randomness, not uncertainties associated with a lack of information.

Traditional Statistical approach starts with a series of predetermined equations (sets of hypotheses) such as single or multivariable linear regression or nonlinear regression that is well defined (e.g., logarithmic, exponential, quadratic). Then, it tries to find the most appropriate, predetermined equation that would fit the collected data. While the Traditional Statistical approach tries to find correlations between the involved variables, it does not address, consider, or even try to identify causations from the data. Everyone knows that correlation does not necessarily address causation. Figures 5.1 and 5.2 are good examples of such an issue. In these two figures, while the data look incredibly correlated (more than 99% correlation) between two topics ("U.S. Spending on science, space, and technology" versus "Suicide by hanging, strangulation, and suffocation" in Figure 5.1 and "Divorce rate in the state of

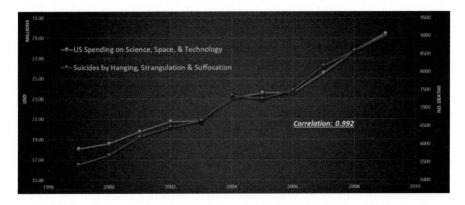

FIGURE 5.1   The correlation between the amount of U.S. spending on science, space, and technology and the number of suicides by hanging, strangulation, and suffocation clearly does not imply causation.

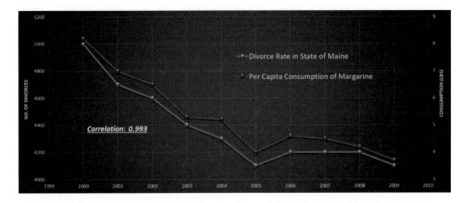

FIGURE 5.2   Neither is causation implied by the correlation between the divorce rate in state of Maine and the per capita consumption of margarine.

Maine" versus "Per capita consumption of Margarine" in Figure 5.2), it is quite clear that these two topics have absolutely nothing to do with one another. This means that correlations do not necessarily mean causation.

Artificial Intelligence does not start with any predetermined models or equations. They do not start with any assumptions regarding the type of behavior that variables may have to correlate them to the target output. The characteristics of Artificial Intelligence and Machine Learning are to discover patterns from existing data. The strength of open Artificial Intelligence using Machine Learning Algorithms has to do with their amazing capabilities to discover highly complex patterns within large amounts of variables. The final outcome of the models that are developed by Artificial Intelligence using Machine Learning Algorithms usually cannot be summarized by one or a few equations. That is why many engineers like to use the term "black box" for the Machine Learning Algorithms that they use to model and solve engineering-related problems.

Apparently, traditional engineers cannot make sense of models unless they include one or more well-defined mathematical equations that they have grown up with since they started college. Most probably, that is the reason behind calling the Artificial Intelligence-related models black box models. What the traditional engineers who do not respect new technologies seem not to realize is that, from a mathematical point of view, the models that are developed using Artificial Intelligence include a series of matrices that generate the models' outcomes. Therefore, nothing is opaque (black) about these models (boxes). The fact is that, when you ask a question from an AI-based model (as long as it has been developed correctly), it can respond in a very solid manner. This part of the topic of Artificial Intelligence is covered in Chapter 7 of this book.

Figure 5.3 summarizes one of the major differences between Traditional Statistical approach to problem-solving and the approach that is used by Artificial Intelligence. When data is used in the context of Traditional Statistical approaches, a series of predetermined equations are used in order to find out how they would possibly fit the data. In other words, the form and shape of the equation that will fit the dataset, and sometimes even the form of the data distributions, are all predetermined. In such cases, if the actual data is too complex to fit any predetermined equations or distributions, then either the approach to solving the problem comes to a standstill or a large number of assumptions, simplifications, and biases are incorporated into the data (or the problem) to come to some sort of solution.

When it comes to modeling the physics of engineering-related problems from data using Artificial Intelligence no pre-determined set of equations is used to fit the data. The main reason for this separation is the avoidance of biases, pre-conceived notions, gross assumptions, and problem simplifications when Artificial Intelligence is used in order to model and find solutions for the physics of engineering-related problems. When Artificial Intelligence is used for engineering modeling of physical

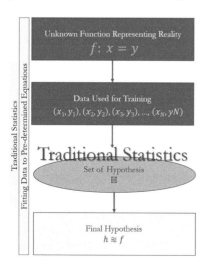
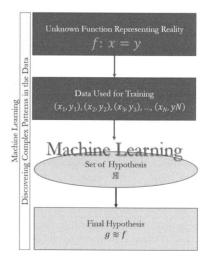

FIGURE 5.3 Differences between Traditional Statistics and Artificial Intelligence & Machine Learning [14].

phenomena, its model and solution are all about the actual data. Data, which in the industry is, in most cases, all about actual (field) measurements, must guide the solutions that are reached, not our today's understanding of the physics that is modeled using mathematical equations. The fact is that the actual (field) measurements represent the physics of the problem that is being modeled. Therefore, by using the data that represents physics, this approach becomes a physics-based problem-solving. The data represents the physics that is being modeled.

Figure 5.4 shows a summary of some of the algorithms that are used in Traditional Statistical approaches versus some of the Machine Learning Algorithms that are used in Artificial Intelligence. While in some books the experts of Artificial Intelligence have shown that Machine Learning Algorithms are able to perform traditional algorithms (example: Linear Regression, Non-Linear Regression, …), some engineers start calling those traditional algorithms, Machine Learning.

Another difference between Traditional Statistics and Artificial Intelligence has to do with the incorporation of deductive reasoning by Traditional Statistics and the incorporation of inductive reasoning by Artificial Intelligence. Deductive reasoning is the process of drawing conclusions based on premises that are generally assumed to be true, while inductive reasoning moves from specific observation to broader generalizations and theories. Deductive reasoning learns by following an intuitive pathway from general to specific (the "*detail*" people), while inductive reasoning learns by following an intuitive pathway from specific to general (the "*big picture*" people).

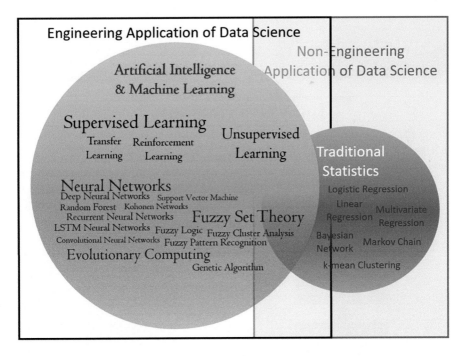

FIGURE 5.4   Different algorithms used by Traditional Statistics versus Artificial Intelligence & Machine Learning.

While both Traditional Statistics and Artificial Intelligence operate on patterns of the provided data, Traditional Statistics uses predefined patterns (i.e., Parametric Models) and compares some measure of the observations to the standard metric of the models (*testing hypotheses*), while Artificial Intelligence builds a model with the data and does not start with an existing model. Traditional Statistics uses an existing model to characterize a pattern in the data. It matches the pre-determined patterns of existing models to the data deductively, following an Aristotelian approach to the truth. Artificial Intelligence uses the patterns in the data to build a model. It discovers patterns in the data inductively. As shown in Figure 5.5, Traditional Statistics uses deductive reasoning moving from general principles to special cases, while Artificial Intelligence uses inductive reasoning, moving from special cases to general principles.

As shown in Figures 5.6 and 5.7, both Traditional Statistics and Artificial Intelligence operate on patterns. While Traditional Statistics uses a predefined pattern (i.e., the Parametric Model) and compares some measure of the observation to the standard metric of the model (*testing hypothesis*), Artificial Intelligence builds a model with the data, and it does not start with a model. Traditional Statistics uses

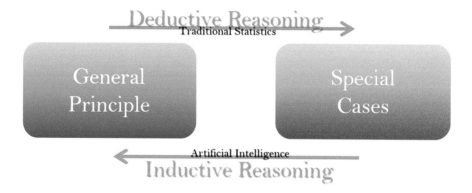

FIGURE 5.5 Traditional statistics uses deductive reasoning while Artificial Intelligence uses inductive reasoning.

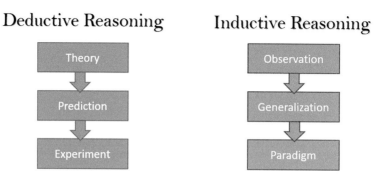

FIGURE 5.6 Differences between deductive reasoning and inductive reasoning.

Deductive Reasoning

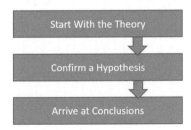

Inductive Reasoning

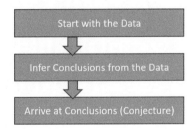

FIGURE 5.7 Differences between deductive reasoning and inductive reasoning using the available data processing.

past information to determine the future state of the system (predict). Artificial Intelligence uses past information to construct patterns based not solely on the input data but also on the logical consequences of those data. Artificial Intelligence predictions contain a vital element that is missing in Traditional Statistics. The ability to provide an orderly expression of what might be in the future, compared to what was in the past (based on the assumptions of the statistical methods), and compared to Statistics that are often hindsight, Artificial Intelligence finds patterns that predict the future.

Artificial Intelligence provides a more complete understanding of data by finding patterns previously not seen. It makes models that predict, thus enabling people to make better decisions, take action, and therefore mold future events.

6 Ethics of Artificial Intelligence (AI-Ethics) in Science and Engineering

It is a well-known fact that predictive modeling and solutions developed by Artificial Intelligence and machine learning algorithms are based on "Data". Since it is known how data is used to build models using Artificial Intelligence, the main characteristics of AI-Ethics are about addressing how AI models become biased or being told (and characterized) how to generate results based on the quality and quantity of the data that is used during AI-based model development.

When it comes to the non-engineering application of Artificial Intelligence and machine learning, it has been proven that human biases such as racism and sexism can be included in the AI models through the inclusion of biased data during the training of the machine learning algorithms. Since science and engineering applications of Artificial Intelligence and machine learning are used to model physical phenomena, AI-Ethics can determine and clarify how human biases, or their concepts and ideas (meaning they want the type of results that should be generated by the AI-based model), can be used in the science and engineering applications of Artificial Intelligence. This is due to the fact that traditional approaches to solving engineering-related problems usually include assumptions, interpretations, and simplifications, and in some cases, they can also provide the opportunity to include preconceived notions and biases.

INTRODUCTION OF AI-ETHICS

The main reason that nuclear weapons did not end up destroying our planet (at least till now) had to do with the worldwide treaties and agreements on how to handle nuclear bombs. It is important that a similar set of worldwide treaties and agreements be eventually achieved by politicians around the world about Artificial Intelligence. One of the main reasons that many individuals are worried about how Artificial Intelligence is going to impact our world in the next few decades has to do with the governments of several countries. Governments in some countries are using this technology based on their own objectives that are a function of their views, beliefs, and understanding or hatred of democracy, as well as their intention of becoming the world leader based on how Artificial Intelligence can serve them. The Ethics of Artificial Intelligence (AI-Ethics) has lately become an important topic that must be well understood by individuals that already are, or are currently becoming, interested in Artificial Intelligence and machine learning algorithms. Unfortunately, a large number of scientists and engineers are not paying any attention to AI-Ethics

DOI: 10.1201/9781003369356-7

since they believe that solving science and engineering related problem would not be impacted by AI-Ethics, which is not true at all, and this is the main reason for this chapter.

Since the mid-2000s, when AI-based image recognition, voice recognition, facial recognition, object recognition, and autonomous vehicles were exposed to most people around the world, interest in Artificial Intelligence has increased significantly. As a new science and technology, Artificial Intelligence will change a lot of things in the 21st century. Artificial Intelligence has become one of the most interesting technologies that people, companies, and academia are getting involved with on a regular basis.

For example, in the past decade, several banks have started using Artificial Intelligence-related models (developed by some of the best-known AI modeling companies) as the first step in making decisions about providing loans to applicants. Since usually hundreds or thousands of individuals apply for loans in different banks, and since each of the banks has only a certain number of employees and their job is to study and analyze the characteristics of loan applicants, it is quite tough to go through the details of hundreds or thousands of individuals that have been applying for loans. In that context, it did make sense for several banks to purchase models developed through Artificial Intelligence to identify the top few loan applicants, for example, 20–50 individuals. Then the bank employees would be able to cover all the details of these top applicants to provide them with bank loans.

The same thing has been true about hiring employees by several companies. This is due to the fact that many companies have a number of human resources department that would be able to analyze and make reasonable decisions about hiring only a few employees each day, week, or month. When some of these companies generate advertisements to hire 10 or 20 new employees, sometimes a very large number of people may apply for such employment. This was the reason that some companies purchased models developed through Artificial Intelligence to identify the top few employee applicants so that the company's human resources employees could analyze and make reasonable decisions about hiring the individuals who had applied for the jobs that had been advertised.

Several companies in the past decade have developed Artificial Intelligence models that could have helped banks decide and make decisions about whom to provide the loans and companies decide and make decisions about who to hire. Banks use AI models to minimize the number of loan applicants, so bank employees must evaluate their characteristics in much more detail. Also, companies use AI-based models to evaluate the large number of applicants that have applied for employment based on the company's job advertisement and then significantly reduce the number of applicants that the actual human resources professionals must concentrate on. Originally, in the past decade, the way Artificial Intelligence has been used by banks and companies to provide loans or hire individuals has made AI-Ethics an incredibly important topic to understood. These are only two examples of many similar things that happened during the last decade.

AI-Ethics is also incredibly important to scientists and engineers who have become enthusiasts of using this technology for solving science- and engineering-related problems. In the past several years, AI-Ethics has become an important topic in the application of Artificial Intelligence in many non-scientific- and non-engineering-related items. This chapter will cover and explain the importance of AI-Ethics in the science and engineering applications of this technology. To clarify the science- and engineering-related AI-Ethics, some examples will also be presented in this chapter.

When there is a lack of understanding of AI-Ethics by scientists and engineers that their effort is to solve physics-based problems using Artificial Intelligence, certain types of problems are created (intentionally or un-intentionally). The main reason for creating such problems has to do with not paying attention to their understanding of AI-Ethics. Without giving full attention to AI-Ethics, many problems are created through the application of Artificial Intelligence to solve science- and engineering-related problems. The impacts of AI-Ethics in such situations usually have to do with

a. lack of scientific understanding of Artificial Intelligence,
b. lack of success in realistic problem solving of physics-based issues through science and engineering application of Artificial Intelligence, or
c. incorporation of traditional engineering biases (assumptions, interpretations, simplifications, and preconceived notions) into the AI-based models of physical phenomena.

Currently, some individuals and companies that claim they use engineering applications of Artificial Intelligence are including a large amount of human bias in problem solving so that they can solve problems using machine learning algorithms after they fail to build an AI-based model that is based on facts, reality, and actual data and does not include human biases. Human biases in engineering have much to do with how mathematical equations are built to solve physics-based problems.

DATA IS THE FOUNDATION OF AI-BASED MODELING

Artificial Intelligence uses machine learning algorithms to develop tools and models to accomplish its objectives. The development of Artificial Intelligence models has a lot to do with "Data". The quality and quantity of the data have a major impact on how the AI-based model will behave. As mentioned in the last section of this chapter, banks have started using Artificial Intelligence models to make the first step in the decision-making about giving loans to applicants. The AI models are usually developed using historical data provided by the loan applicants along with previous results of loan payments. The amount of positive and negative loan payments as well as the input data from the loan applicants, such as gender, ethnicity, credit, living location, income, etc., will determine the quality of the AI-based model that is developed for the bank loan. Such models can also include certain characteristics determined by the bank's management.

The same general approach is also applicable to the AI models for the human resources of large companies to make the decision about who to hire. Such models are also developed using existing data from multiple companies about employee applicants as well as the quality of the employees that have been hired in the past. Other applications of AI that make AI-Ethics highly important include Face Recognition, Face Detection, Face Clustering, Face Capture, Face Match, etc. Such technologies are used by mobile phones, security, police, airports, etc.

In the science and engineering application of Artificial Intelligence, the characteristics of the data (including its quality and quantity) that is used for model development impact the quality of the AI-based models. Science and engineering application of Artificial Intelligence is the use of actual measurements and actual physics-based data to model physics rather than using mathematical equations to build models for physical phenomena. Traditionally, in the past few centuries, modeling physics at any given time had to do with engineers' and scientists' understanding of the physical phenomena that were being modeled. As scientists' understanding of any physical phenomenon enhances, so do the characteristics of the mathematical equations that are used to model that physical phenomenon.

Data that are used by non-scientific and non-engineering-related problems, such as the examples that were mentioned in the previous sections of this chapter (Loans from Banks and Employee Hiring by Companies), are usually generated by humans. The data generated by humans (their activities that are saved as data to be used) is not based on the facts and realities that are measured. Such data has much to do with the behavior of the humans that generate it. Such data has to do with decision-making by humans that may have certain types of behavior, understanding, intelligence, or even political issues.

AI-ETHICS ADDRESSES THE BIAS IN AI-BASED MODELING

The characteristics of the quality and quantity of the data that is used to build the AI-based model determine whether any biases have been incorporated in the model developed by Artificial Intelligence. AI-Ethics' objective is to identify the quality and quantity of the data that is used to build the AI-based model and to identify if any bias has been (intentionally or unintentionally) incorporated in the model through the data that has been used to build the model.

The way Artificial Intelligence and machine learning have been used by banks and companies to loan or hire individuals has made AI-Ethics an incredibly important topic to understand. The same is true about the science and engineering applications of Artificial Intelligence and machine learning. As long as realistic and non-traditional statistics and machine learning algorithms are incorporated, the quality of the AI-based models is purely based on the quality and quantity of the data that has been used to build the model. Therefore, the data that is used to develop the AI-based model completely controls the essence of the model that is developed and used for decision-making.

As Artificial Intelligence moved forward and started solving more problems, scientists became interested in learning more details about how Artificial Intelligence works. It becomes quite clear that the main characteristic of Artificial Intelligence is its use of data to come up with the required solutions and to make decisions. Since data is the main source of AI-based model development, it became important to learn:

A. where the data is coming from and what the main source of it is, and
B. to what extent the data includes all the required information (even not explicitly) that Artificial Intelligence can extract patterns, trends, and information from.

For example, a company was developed in 1975 and in its first 25 years (from 1975 to 2000), its management hired all its employees based on their intellectual and quality rather than considering any racism and/or sexism. However, the company went through a management modification in the year 2000. The new management that was took over the same company starting in 2000 was just as important about employees intellectual and quality, but it was also very important to them the type of employees that would be hired (Figure 6.1). This had to do with the company's management's racism and/or sexism. Let's assume that this is a fact and will create a specific type of data about hiring employees in the past several years.

When it comes to another new management for this same company in 2015, they use an AI company to build them an AI-based system (AI-based model) that would help them hire new employees. The AI company will ask the management to provide them with data from their past several years of employment hiring. Depending on to what degree the previous several years of employee hiring data is provided to the AI company, they will be using it to train an AI-based system (AI-based model) to help them hire new employees. If only the data from the last 15–20 years is provided to the AI company, then the AI-based system (AI-based model) for hiring new employees would be a racist and sexist AI model because the data that was used to develop the model has learned from the biases that were included in the data used to train and develop it.

It took almost a decade of research and study until it became quite clear through examining the actual application of this technology that AI and machine learning have the potential to be political [15,16], racist [17,18], and sexist [18,19]. This has to do with the type of data that is used to build the AI and machine learning models. In other words, it is quite possible to create a biased AI and machine learning model that can do what you want it to do. It completely has to do with the data that is used to train and build the model. This is how AI-Ethics addresses the engineering application of AI when traditional engineers intentionally (or unintentionally) modify the quality of the AI-based models so that they generate what they believe

FIGURE 6.1   In the past 25 years, this company has had two series of managements. The first company management in the first 25 years was not sexist and racist, but in the last 15 years, the company included sexist and racist management.

is the right thing rather than modeling the physical phenomena based on reality, fact, and actual measurements.

MIT's AI-Ethics has published articles regarding the biases that can take place when using AI and machine learning. In some of these articles, it is clearly mentioned that "Three new studies propose ways to make algorithms better at identifying people in different demographic groups. But without regulation, that won't curb the technology's potential for abuse" [20], and "this is how AI bias really happens—and why it's so hard to fix. Bias can creep in at many stages of the deep-learning process, and the standard practices in computer science aren't designed to detect it" [21].

In another interesting article, it is mentioned that "Collecting the data; There are two main ways that bias shows up in training data: either the data you collect is unrepresentative of reality, or it reflects existing prejudices. The first case might occur, for example, if a deep-learning algorithm is fed more photos of light-skinned faces than dark-skinned faces. The resulting face recognition system would inevitably be worse at recognizing darker-skinned faces. The second case is precisely what happened when Amazon discovered that its internal recruiting tool was dismissing female candidates. Because it was trained on historical hiring decisions that favored men over women, it learned to do the same" [22]. What has been mentioned in this article are the results of research that was done to learn how bias can be included in the model. This is so true and very important for both engineering and non-engineering applications of AI and machine learning. In this chapter, it will be shown how similar activities take place in the engineering application of AI, and it will be explained in the next section what bias is when AI is used to model physical phenomena.

By doing some serious research on the fundamentals of Artificial Intelligence and machine learning algorithms, it becomes quite clear that this technology has an incredibly strong power to discover patterns in the data that is used to train and develop models, make predictions, and help with decision-making. Since what Artificial Intelligence and machine learning algorithms do is all about data, it also becomes clear that as long as the data that is provided to them is generated based on biases, interpretations, and assumptions, the models and workflows that this technology develops become representative of such biases, interpretations, and assumptions.

APPLICATION OF AI-ETHICS IN SCIENCE AND ENGINEERING

Given the experience of how Artificial Intelligence was used in many situations that created racism and sexism, it is quite clear that it is quite possible to control the behavior of the Artificial Intelligence-related models and systems. Since this has been proven in general intelligence-related situations, many scientists and engineers think it has nothing to do with modeling physical phenomena. Actually, an overwhelming number of scientists and engineers that have become interested in Artificial Intelligence never pay any attention to AI-Ethics because they believe that such issues are not applicable to science and engineering. This is not true. When any scientist or engineer comes to such a conclusion, it means that they have very little understanding of Artificial Intelligence. While racism and sexism are not applicable to science and engineering (the application is not working in the industries), they are the results of "Bias" that are included in the data that is used to build the models and systems using Artificial Intelligence. In such cases, AI-Ethics is completely applicable to science and engineering for creating certain types of biases to control Artificial Intelligence-based models and systems.

AI-Ethics is important to engineers and scientists who have become enthusiasts of using this technology for solving engineering-related problems. While in the past several years AI-Ethics has become an important topic in the non-engineering application of Artificial Intelligence and machine learning, now it is just as important in the science and engineering application of this technology. Examples of AI-Ethics in the engineering application of AI and machine learning are presented in this chapter, showing how guesswork, assumptions, interpretations, and simplifications can help traditional engineers use AI and machine learning algorithms to generate an unrealistic and highly biased predictive model.

This usually happens when they are not successful in using the actual field measurements for building AI-based models of physical phenomena. What they do not realize is their lack of understanding and realistic knowledge of Artificial Intelligence. Therefore, to make their claim of building AI-based modeling work (usually they call it "Hybrid Models", or "Physics-Based AI Model") they start combining data generated from mathematical equations with actual field measurements. What they do not realize (or do realize but do not mention it to those they make the model for) is that the data generated from mathematical equations already includes correlations and patterns between all the variables and parameters that have been used in the mathematical equations. Since the major characteristics of the machine learning algorithms for Artificial Intelligence try to perform pattern recognition, the patterns on the data that are generated by the mathematical equations tell the machine learning algorithm what to do.

It seems that the reasons behind the inclusion of such biases in the engineering application of Artificial Intelligence have much to do with the lack of scientific understanding of how Artificial Intelligence must be used to model physical phenomena. Details of how to model physical phenomena using Artificial Intelligence are covered in Chapter 4 of this book. Currently, a large number of individuals and companies that claim they are using Artificial Intelligence to solve science- and engineering-related problems are including a large amount of human bias in their technology to be able to generate solutions. The main reason for such an approach is the fact that they fail to use actual measurements (real data) to build AI-based models of physical phenomena that do not include human biases. Human biases in engineering have much to do with how mathematical equations are built to solve physics-based problems.

The major contribution of Artificial Intelligence and machine learning to science and engineering is problem solving and modeling physical phenomena based on actual measured data, which would be the main core behind the avoidance of biases, assumptions, interpretations, and preconceived notions about physics. Since traditional techniques for modeling physical phenomena are through mathematical equations, they usually include assumptions and sometimes provide the opportunity to include biases. This is very true when the physical phenomena that are being modeled cannot be seen, looked at, or even touched, such as petroleum engineering, which is a good example of such a situation since the produced hydrocarbon is a deeply underground fluid. The same is true about any other engineering discipline when mathematical equations that include assumptions, interpretations, and simplification are used to model the physical phenomena.

Reservoir engineering, reservoir modeling, and reservoir management overwhelmingly contribute to most of the income of the operating and service companies in the oil and gas industry. This shows why reservoir modeling is a very important technology in the petroleum industry. Currently, the same is true about climate change when CO_2 is stored in geological formations. It is a fact that modeling fluid flow in hydrocarbon reservoirs includes serious amounts of assumptions, interpretations, and simplifications since they are hundreds or thousands of feet below the surface. This means that it has been impossible to observe, touch, or realistically test anything that takes place in a hydrocarbon reservoir. In this section, let's see how AI-Ethics is being significantly navigated in petroleum engineering by those that claim to be using Artificial Intelligence.

Figures 6.2 and 6.3 show examples of actual measured data from conventional and unconventional resources that are being used to produce oil and gas. These plots clearly show that the actual field measurements that are collected hardly show any correlations with the oil and gas productions that are the main characteristics of the petroleum companies throughout the world. When the complexities of the actual measured data are used to model such physical phenomena using Artificial Intelligence and machine learning, the overwhelming majority of the vendors and service companies in the oil industry end up not being able to develop AI-systems and AI-models for such complex physical processes.

Since such marketing companies are unable to achieve their AI-based solutions using complex, actual measured data, they have come to the conclusion of generating what they currently call "Hybrid Models". The definition of "Hybrid Models" is the combination of "Actual Field Measurements" with "Data Generated through Mathematical Equations". It is a well-known fact that all data generated through mathematical equations include certain types of correlations between all the parameters (variables) that are included in the mathematical equation. When the large amount of data already includes well-known correlations, then such correlations are going to generate the pattern recognitions that are the main characteristics of the machine learning algorithms from the provided data. Since the overwhelming majority of mathematical equations already include assumptions, interpretations, and simplifications, their data generations that become included in the "Hybrid Models" will determine what kind and type of pattern recognitions must be generated by the machine learning algorithms. This means that using "Hybrid Models" we are already telling Artificial Intelligence what kinds of solutions must be generated, rather than providing only actual measured data and asking Artificial Intelligence to generate the type of pattern from the actual data to find the reality of the physical phenomena that are being solved. The type of mathematical equations that are used to generate the data that is hybridized with the actual field measurements tells the machine learning algorithm what to do. This is the "Bias" that is the main issue in AI-Ethics.

Here's an example of what was mentioned in the above paragraph. Figure 6.4 provides an example of a series of actual field measurements from two (out of several)

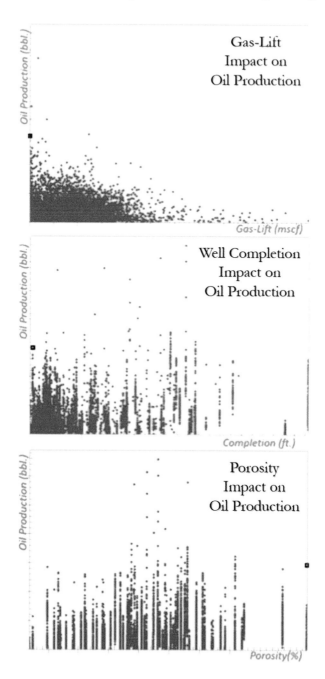

FIGURE 6.2 Actual measured data from a field in Southeast Asia. All three plots show oil production versus gas lift (on the left), completion (in the middle), and porosity (on the right).

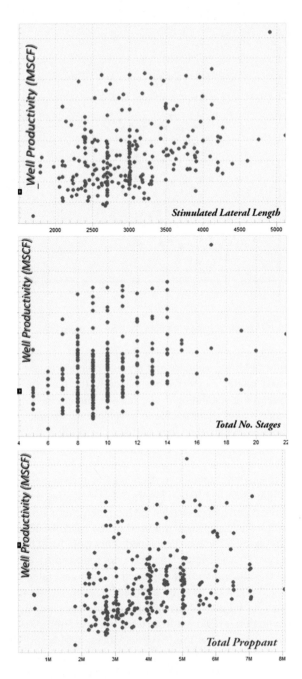

FIGURE 6.3 Actual measured data from a Marcellus Shale field in Southwest Pennsylvania. All three plots show gas production versus stimulated lateral length (on the left), total number of stages (in the middle), and total amount of proppants (on the right).

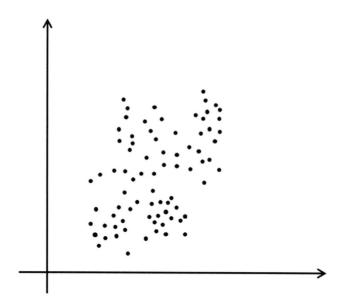

FIGURE 6.4   Example of actual measured data from an example field.

of the variables in a field that vendors and service companies are trying to use to develop an AI-based model. Let's assume that they try hard for model development, and their limited understanding of the science and engineering application of Artificial Intelligence makes them unable to develop a reasonable AI-based model. Once they are unable to do that, then they will start working with "Hybrid Models". One of the companies uses a specific mathematical equation that can generate data for several variables.

Data generated through the specific mathematical equation used by this first company is shown in Figure 6.5. Just like in Figure 6.4, the generated data is only shown for these two specific parameters (variables). Once the generation of such data from mathematical equations is completed and used in the machine learning algorithm, the model shown in Figure 6.6 can be developed. This would still be called AI-based by such companies as "Hybrid Model" that have not followed AI-Ethics.

Figures 6.7 and 6.8 show a similar approach to data generated through another specific mathematical equation used by another company, while the original data measurements shown in Figure 6.4 are being used by this company as well. Since the mathematical equation is different from what was used and shown in Figure 6.5, the generated AI-based model for this "Hybrid Model" will be different from what is shown in Figure 6.6. "Hybrid Model" developed by the second company from the same actual data measurements (Figure 6.8) is different from the "Hybrid Model" developed by the first company that has used the same actual data measurements (Figure 6.6).

Figures 6.9 and 6.10 show how different biased "Hybrid Models" can be developed when scientists and engineers have no understanding of AI-Ethics.

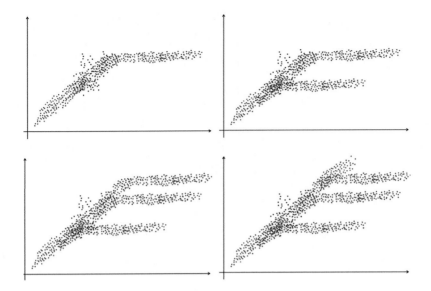

FIGURE 6.5   Example of data generated by using the mathematical equation for several variables. This shows plots of two variables.

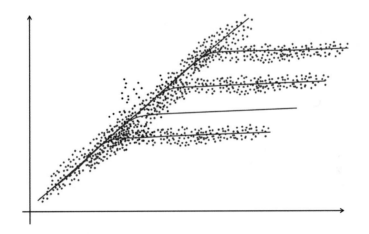

FIGURE 6.6   The model that can be generated using the "hybrid model" identifying the model characteristics through data generated by mathematical equations.

FIGURE 6.7   Example of data generated by using <u>another</u> mathematical equation for several variables. This shows plots of two variables.

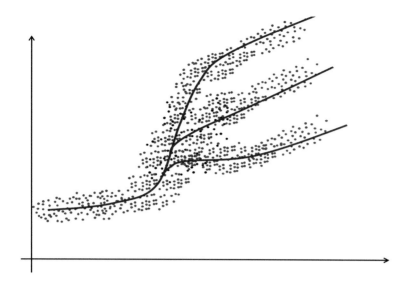

FIGURE 6.8 The model that can be generated using the "hybrid model" identifying the model characteristics through data generated by <u>another</u> mathematical equation.

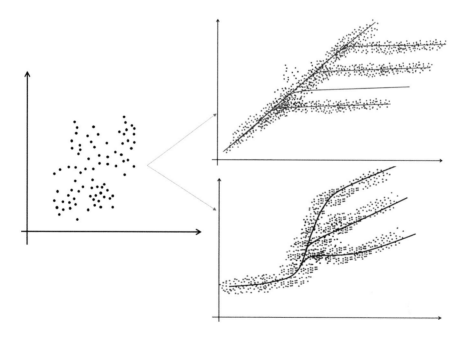

FIGURE 6.9 Actual measured data (on the left) is combined with generated data from two different mathematical equations shown on the right.

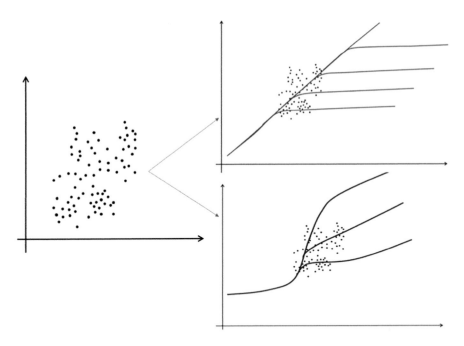

FIGURE 6.10   Through bias identified by AI-Ethics, the actual measured data (on the left) is used to develop two different AI-based models using the "hybrid model" approach.

Since I am a petroleum engineering scientist, I provide examples related to petroleum engineering as well and try to explain them in the best fashion to make sense to all scientists and engineers. In the past decade, the overwhelming majority of hydrocarbon (oil and gas) production in the United States of America has been produced from shale reservoirs. Shale is known as an unconventional reservoir since it is very different from conventional reservoirs such as sandstone and carbonate reservoirs. All the reservoir technologies were originally developed for conventional reservoirs, but now the conventional technologies are being used for unconventional reservoirs such as Shale. This provides far more assumptions, simplifications, and interpretations for dealing with Shale reservoirs when compared with conventional reservoirs. This is an important example of including biases when some people, vendors, and service companies claim that they are using Artificial Intelligence to solve shale-related problems when they use even conventional mathematical equations to generate data that they use to generate models using machine learning algorithms.

It is important to note that some part of the reservoir rock is usually brought to the surface, and it is tested and analyzed in the laboratory to help scientists and engineers develop an understanding of the complexity of the physics of fluid flow in the porous media deep underground. However, realistic facts about such analyses must not be overlooked. While hydrocarbon reservoirs have approximate volumes of hundreds of millions to tens of billions of cubic feet, the part of the rock that is brought to the laboratory for observation and testing is usually less than a few cubic feet. This provides information about what percent of the reservoir will be viewed and tested. Furthermore, it is a well-known fact that hydrocarbon reservoirs are highly heterogeneous, which means what is analyzed in the laboratory over a few square inches of rock is not realistically representing what happens throughout the entire reservoir rock.

Therefore, while laboratory core analysis is an important and useful process for understanding the fluid flow in the hydrocarbon reservoir, it cannot realistically represent all the details and the heterogeneity that happen throughout the tens of billions of cubic feet of the hydrocarbon reservoir, which is hundreds or thousands of feet under the ground. This clarifies the existence of assumptions, interpretations, and simplifications in the mathematical equations that are used to model the fluid flow in porous media. Furthermore, when the hydrocarbon reservoir is unconventional, such as shale plays, which are currently the main source of hydrocarbon production in the United States, the problem mentioned above becomes orders of magnitude more complex and even more important.

The number of assumptions, interpretations, simplifications, preconceived notions, and biases in modeling the physics of completion, hydraulic fracturing, and fluid flow in shale plays is so extensive that the mathematical modeling for completion and production optimization from shale wells is completely unrealistic, useless, and full of bias. This is due to the fact that historical details for understanding the physics of fluid flow in hydrocarbon reservoirs are mainly applicable to conventional plays. This technology that has been developed for conventional reservoirs has been extrapolated to unconventional reservoirs starting a decade ago.

It is a clear fact that the mathematical equations that are used to model the physics of hydrocarbon production from shale wells are overwhelmed by assumptions and hardly have anything to do with facts and realities since the main essence of this technology is mainly applicable to the conventional reservoir and not to the unconventional reservoir. It is hard to find any real scientists and professional engineers (including those that have developed and are using these techniques) to claim that the current version of mathematical modeling of hydraulic fracturing of shale wells has anything to do with reality.

These facts prove that using Artificial Intelligence for the development of so-called "hybrid models" is full of assumptions, interpretations, and biases and has nothing to do with the reality of science and engineering applications of Artificial Intelligence. When such mathematical equations are used to generate data and then combine such data with actual field developments to build so-called "hybrid models", such models can be forced to generate the type of outputs and results that are pre-determined by those that develop them. It removes the actual and real characteristics of Artificial Intelligence that are capable of modeling physics based on reality rather than based on guesswork and biases. This is a good example of how AI-Ethics must be addressed in the science and engineering applications of this technology.

It is a well-known fact that when hydraulic fracturing is performed on unconventional reservoirs that are naturally fractured (such as shale), the results are quite different from what happens when hydraulic fracturing is performed on conventional reservoirs (sandstones). In shale, due to the existence of complex natural fractures, hydraulic fracture creates a "network of fractures" (as shown in Figures 6.11 and 6.12), not an elliptical hydraulic fracture (as shown in Figure 6.2 for conventional reservoirs).

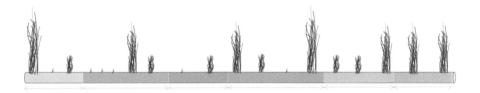

FIGURE 6.11   Hydraulic fracturing of naturally fractured reservoirs cannot be modeled based on real measurements since there is no way to identify the distribution of natural fractures thousands of feet under the ground.

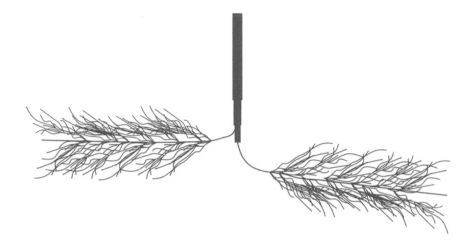

FIGURE 6.12   Hydraulic fracture network in a naturally fractured reservoir.

As shown in Figures 6.11 and 6.12, when liquid (water) is injected in an unconventional reservoir for hydraulic fracturing purposes, prior to the injection of the proppant, it starts to fracture the formation. As the formation starts fracturing, the continuation of the fracture will go through the least resistant pathways in the rock. In a naturally fractured reservoir, the least resistant pathway is the network of the natural fractures, while the actual fabric of the rock (that has not been naturally fractured millions of years ago) has more resistance. Therefore, hydraulic fracturing of unconventional resources such as shale that are naturally fractured reservoirs creates highly complex networks of natural fractures that cannot be modeled in detail. This is due to the fact that the shape, characteristics, and details of the natural fracture of the rock (shale) cannot be observed or measured throughout the reservoir. The highly complex shape of the hydraulic fracture network in the unconventional reservoir is a function of heterogeneity and the natural fracture network.

When the model development of hydraulic fracturing was performed more than 50 years ago, the shape of the hydraulic fracture in the conventional reservoir was modeled using an elliptical shape, as shown in Figure 6.13. This traditional hydraulic fracture model includes four specific characteristics that allow it to be modeled using mathematical equations. These four specific characteristics are (a) fracture half-length,

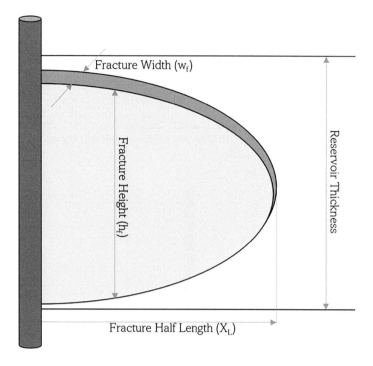

FIGURE 6.13   Hydraulic fracture in a conventional reservoir.

(b) fracture height, (c) fracture width, and (d) fracture conductivity. Comparing the shapes of the hydraulic fractures that are demonstrated in Figures 6.11 and 6.12 versus the shapes of the hydraulic fracture that are shown in Figure 6.13 makes it quite clear how different the actual shape of hydraulic fracture is between unconventional versus conventional reservoirs.

When the actual hydraulic fractures look like what is shown in Figures 6.11 and 6.12, does it make any sense, or does it have anything to do with reality, to model it using the shape that is shown in Figure 6.13? The answer to this question must be quite clear. This is a good example of how assumptions, interpretations, preconceived notions, simplifications, and biases that are included in the mathematical equations (that are used to model the physics of fluid flow in porous media) are included in "hybrid models" that combine them with real field measurements.

AI-Ethics have proven to be an important issue when AI-based models are used in decision-making. AI-Ethics can expose biases that may have been included in the Artificial Intelligence applications of science and engineering as well as general intelligence problem-solving. Several studies, some of which were referenced in this chapter, have shown how biases such as racism and sexism have been included in the AI-based models that have been exposed through AI-Ethics. This chapter demonstrated how assumptions, interpretations, and biases developed by traditional engineers can be included in the engineering application of Artificial Intelligence, which is referred to as AI-Ethics in Engineering.

7 Explainable Artificial Intelligence (XAI)

Explainable Artificial Intelligence (XAI) is one of the main characteristic of Science and Engineering Application of Artificial Intelligence. As it was mentioned in the previous chapters, Science and Engineering Application of Artificial Intelligence models physical phenomena purely based on facts and reality (actual measurement data). The main objective of this new technology, which is quite different from traditional engineering applications, is the complete avoidance of assumptions, interpretations, simplifications, preconceived notions, and biases. One of the major characteristics of Artificial Intelligence for Science and Engineering Application that is different from Artificial General Intelligence is its incorporation of XAI. While using actual data measurements as the main building blocks of modeling physical phenomena, Science and Engineering Application of Artificial Intelligence incorporates several types of Machine Learning Algorithms, including artificial neural networks, fuzzy set theory, and evolutionary computing. Predictive models of Science and Engineering Application of Artificial Intelligence are not represented by the unexplainable "Black Box". Predictive models of Science and Engineering Application of Artificial Intelligence are reasonably explainable.

In the early 1990s, when some engineers started using Artificial Intelligence and Machine Learning to solve engineering-related problems, generate related articles, and present their new technology at conferences, many other engineers and scientists that were being exposed to such models and solutions started asking "how this technology (Artificial Intelligence and Machine Learning) achieves its predictive objectives". What is being addressed recently as XAI for the non-engineering application of Artificial Intelligence, is not new in the context of engineering applications of this technology. The main reason behind the fact that Science and Engineering Application of Artificial Intelligence to a large extent is quite understandable has to do with the historical problems of the application of traditional statistics that were used to solve engineering-related problems.

While traditional statistics' main solutions are about the identification of "correlations" in the data, engineers and scientists were always interested in "causations" that could explain the existing "correlations". This has always been one of the main problems associated with the use of traditional statistics to solve problems. There are still many AI-based solutions that are referred to as "Black Box". It is important to note that Science and Engineering Application of Artificial Intelligence does not generate "Black Box" solutions.

As mentioned before, in the early 1990s, questions by many engineers and scientists about the type of engineering solutions using Artificial Intelligence and Machine Learning gave rise to research and development efforts in early 2000 and resulted in

DOI: 10.1201/9781003369356-8

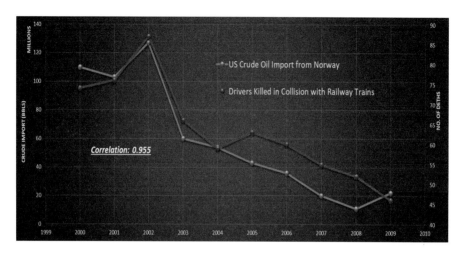

FIGURE 7.1 In 9 years, from 2000 to 2009, U.S. crude oil imports from Norway highly correlates with drivers killed in collisions with railway trains. It is quite clear that such a correlation has absolutely nothing to do with causation.

what is today called XAI. In this chapter, the history of what today is called XAI is shown through seven technical papers in the oil and gas industry that were published between 2001 and 2010, while "Explainable Artificial Intelligence" was brought up as a new item by the U.S. Department of Defense in 2016. Historically, prior to the development of Artificial Intelligence, traditional statistics were used to analyze data. The major role of traditional statistics is to specify hypotheses that fit the collected data. Therefore, the key behind traditional statistics is "correlation" while engineers and scientists have always been interested in "causation". It is a well-established fact that "correlations" do not necessarily determine and/or represent "causations". Causation means that changes in variables interact with one another. Modifications in one variable cause specific types of modifications in another variable.

Figures 7.1 and 7.2 are good examples of how "correlation" does not have anything to do with "causation". These figures demonstrate how data collected for different variables can be correlated to each other while having absolutely nothing to do with one another. Figure 7.1 shows over 95% correlation between "U.S. Crude Oil Import from Norway" and "Drives Killed in Collision with Railway Trains". While 9 years of data from these two items are highly correlated with each other, it is quite obvious that they do not have anything to do with each other. Figure 7.2 shows over 98% correlation between "Ice Cream Sales" and "Shark Attacks" for an entire year. This data also has absolutely nothing to do with each other, while they are highly correlated with each other. These actual data clearly show that correlation does not necessarily explain any causation.

EXPLAINING TRADITIONAL ENGINEERING MODELS

It is a well-known fact that models of physical phenomena that are generated through mathematical equations can be explained. This is one of the main reasons behind

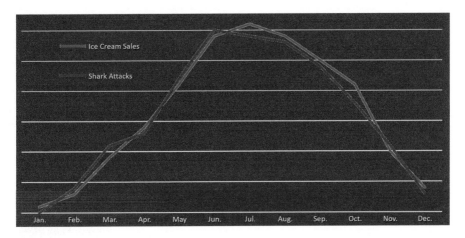

FIGURE 7.2 In the recent 12 months, ice cream sales correlate with shark attacks. Such a correlation has absolutely nothing to do with causation.

the expectation of engineers and scientists that any potential model of the physical phenomena should be explainable. The explainability of the traditional models of physical phenomena is achieved through the solutions of the mathematical equations that are used to develop the models. Explanation of such models is achieved through analytical solutions for reasonably simple mathematical equations or numerical solutions for complex mathematical equations. Solutions to the mathematical equations provide opportunities to get answers to almost any questions that might be asked from the model of the physical phenomena.

Solutions to the mathematical equations are used to explain why and how certain results are generated by the model. It allows examination and explanation of the influence and impact of all the involved parameters (variables) on each other and on the model's results (output parameters). In general, this is the definition of the engineering approach to problem-solving. Therefore, the results of physical phenomena that are based on mathematical equations (no matter how simple or complex) can be explained in detail.

Figure 7.3 shows the Fetkovich "Type Curve". Including early time (driving from transient flow equations) and late time curves (boundary-dominated flow, Arp's Decline Curves). The equations that have been used for this series of "Type Curves" are also provided in Equations 7.1 and 7.2, right after Figure 7.3.

$$q_{Dd} = q_D \left[\ln\left(\frac{r_e}{r_{wa}} \right) - \frac{1}{2} \right] \tag{7.1}$$

$$t_{Dd} = \frac{t_D}{\frac{1}{2}\left[\left(\frac{r_e}{r_{wa}} \right)^2 - 1 \right]\left[\ln\left(\frac{r_e}{r_{wa}} \right) - \frac{1}{2} \right]} \tag{7.2}$$

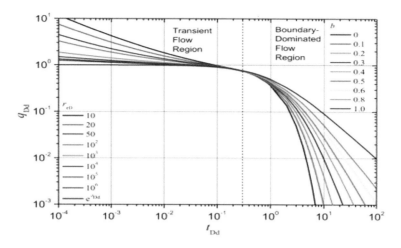

FIGURE 7.3   Fetkovich Type Curves.

Dimensionless rate (q_D) and time (t_D) are used in *Equations 7.1* and *7.2* and are defined as shown below (*Equations 7.3* and *7.4*) in the context of the well-testing domain:

$$q_D = \frac{141.2 \, q \, \mu \, B}{k \, h \left(p_i - p_{wf} \right)} \tag{7.3}$$

$$t_D = \frac{0.00633 \, k \, t}{\varnothing \, \mu \, c_t \, r_w^2} \tag{7.4}$$

It can be clearly seen that the curves in Figure 7.3 are very "well-behaved" (continuous, non-linear, certain shape that changes in a similar fashion from curve to curve). The reason behind the "well-behaved" characteristics of these curves is the mathematical equations that were used to generate them. All "Type Curves" that have been historically generated in the petroleum industry are very "well-behaved" because they are all developed using the solutions of the mathematical equations. Figure 7.4 demonstrates two more examples of such "well-behaved" "Type Curves".

This means that traditional physical phenomenon models that have been developed using mathematical equations are "explainable". Physical phenomena that are developed using mathematical equations are "explainable" through several types of analyses of the mathematical equations. Here are four ways of explaining physical phenomenon models that are developed and solved using mathematical equations:

1. Identification of the influence of each parameter (variable) in the mathematical equation, also known as Key Performance Indicators (KPIs),
2. Single-Parameter Sensitivity Analysis,
3. Multiple-Parameter Sensitivity Analysis, and
4. "Type Curve" Generation.

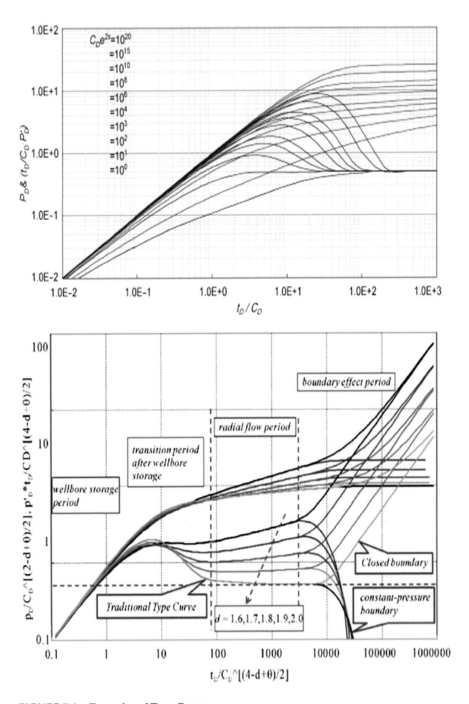

FIGURE 7.4   Examples of Type Curves.

The performance of such analyses can provide the required explanation of any question that might be asked regarding why and how certain outcomes are generated, or certain forecasts and predictions are made based on the solutions of the mathematical equations that have been used to model the physical phenomena. Figures 7.3 and 7.4 demonstrate examples of "Type Curves" that are generated through the solutions of certain mathematical equations.

EXPLAINING MODELS OF PHYSICS DEVELOPED BY ARTIFICIAL INTELLIGENCE (XAI)

In a recent article that was published at Forbes [23], it was mentioned:

"Many of the algorithms used for machine learning are not able to be examined after the fact to understand specifically how and why a decision has been made. This is especially true of the most popular algorithms currently in use – specifically, deep learning neural network approaches. XAI is an emerging field in machine learning that aims to address how black-box decisions of AI systems are made. So far, there is only early, nascent research and work in the area of making deep learning approaches to machine learning explainable".

This article was written on July 23, 2019. It demonstrates that systems and models mimicking human-level intelligence (non-engineering-related problems) that have been developed using Artificial Intelligence and Machine Learning have major issues with explaining how this technology predicts, forecasts, or makes decisions. The types of applications of Artificial Intelligence and Machine Learning that are referenced in this article, as well as many other recent articles about XAI, are mainly related to the application of this technology to non-engineering-related problems. When it comes to the engineering application of Artificial Intelligence and Machine Learning, "how and why a decision has been made" becomes far more important than when this technology is used for non-engineering-related problems. Therefore, what today is being called XAI has been a major issue since this technology started to be used in engineering in the early 1990s.

It is interesting to note that explainable predictive models developed using Artificial Intelligence and Machine Learning were developed by a technical company named "Intelligent Solutions, Inc.".[1] in the early 2000s. These XAI models are part of the application of this technology in Petroleum Data Analytics (PDA)[2]. As was mentioned before, the main reason behind the research and development of what is today called XAI, which took place much earlier, had to do with the fact that once this technology is applied to engineering-related problem-solving, explainability of the models becomes a critical issue for scientists and engineers.

The main reason behind calling the predictive models that are generated using Artificial Intelligence and Machine Learning algorithms "Black Box", has to do with the fact that these algorithms do not model physical phenomena using mathematical equations. When it comes to the engineering application of Artificial Intelligence and Machine Learning, the "Black Box" characteristics of such models will cause serious problems for scientists and engineers. Many traditional engineers who have a negative view of this new technology usually use the term "Black Box" to deny the contribution of this technology to the future of engineering problem-solving.

One of the major contributions of PDA that has been developed during the past three decades at Intelligent Solutions, Inc. and West Virginia University is the creation of transparency for the so-called "Black Box" of Predictive Analytics. Since PDA is a purely physics-based technology that avoids any mathematical equations and generates purely Artificial Intelligence based predictive models, it develops explainable predictive models. The objective of this chapter is to demonstrate how "Explainable the Artificial Intelligence" (XAI) can be provided for engineering-related models that have been developed using Artificial Intelligence. Since the author of this book is a petroleum engineer, the XAI examples that are shown in this chapter would all be related to petroleum engineering. However, in the same way, this can be done for all other engineering and scientific technologies that are modeled using Artificial Intelligence. This demonstration will be explained through KPIs, Sensitivity Analysis, and Type Curves.

PART 1 OF XAI: KEY PERFORMANCE INDICATORS (KPIS)

AI-based engineering models that have been developed in the context of PDA can provide a tornado chart to demonstrate and rank the contribution of all the input parameters that were used to develop (train, calibrate, and validate) the predictive model. What is shown in the next 19 figures in this chapter (Figures 7.11, 7.13–7.25, 7.27 and 7.28, and 7.30–7.32) that have been developed to generate the XAI in engineering models were performed by an engineering software application that is called "IMprove" and has been developed by "Intelligent Solutions, Inc.".

Figure 7.5 shows a tornado chart generated from the Artificial Intelligence engineering-related model developed using the technology that has been explained in detail in a book called "*Shale Analytics*" [24] for a Marcellus Shale field in southwestern Pennsylvania. The Artificial Intelligence predictive model that was developed for this reservoir and completion engineering-related problem included 24 different field measurements. The output of this AI-based predictive model was "30 days of cumulative gas production" from each well in this field.

The KPI tornado chart of this Artificial Intelligence predictive model shows that, on average, the operational condition in this field plays the most important role in controlling the first 30 days of gas production. Then it is followed by attributes associated with Completion Design, Well Characteristics, Hydraulic Fracture Implementation, and finally the Formation Characteristics. In Figure 7.5, there are background colors for each of the attributes (input parameters) on the top part of the figure, and in the bottom of the figure, the background color defines the category of input parameters and also shows the level of importance of the categories of input parameters that are shown at the top of the figure.

Rank	Attribute	% Degree of Influence
1	Stimulated Lateral Length(ft)	100
2	Net Thickness(ft)	98
3	Longitude	81
4	Clean Volume(bbl)	74
5	WPH – 30 days (psi)	70
6	Max. Prop. Concentration(lbs./gal)	66
7	Prop. /Stage(lbs.)	51
8	Latitude	47
9	Avg. ISIP	47
10	Bulk Mudulus	46
11	Total No. of Stages	45
12	Porosity(%)	45
13	Shear Mudulus	45
14	Total No. Clusters	44
15	Avg. Inj. Pressure(psi)	34
16	True Vertical Depth (ft.)	34
17	Shot Density (Shots/ft)	31
18	Slurry Volume(bbls.)	28
19	TOC(%)	27
20	Avg. Inj. Pressure(psi)	23
21	Youngs Modulus	22
22	BTU Area	20
23	Swi(%)	16
24	Poisson's Ratio	1

Rank		% Degree of Influence
1	Operational Conditions	100.00
2	Completion Design	79.00
3	Well Characteristics	77.00
4	Hyd. Fracture Implementation	66.00
5	Formation Characteristics	51.00

FIGURE 7.5 Tornado chart of key performance indicators generated in Shale Analytics (application of petroleum data analytics in unconventional reservoirs) for a data-driven predictive model for a Marcellus Shale field in southwest Pennsylvania.

PART 2 OF XAI: SENSITIVITY ANALYSIS

One of the major techniques that can be used to explain the behavior of a model is sensitivity analysis. The engineering application of Artificial Intelligence (predictive analytic) models that are provided in this chapter is in the context of PDA. These AI models include several input parameters that are used to build an Artificial Intelligence physics-based model purely based on actual field measurement data to generate the required output. In the context of reservoir engineering, let's consider building an AI-based predictive Shale Analytics model that uses reservoir characteristics as well as completion design and implementation (as part of other input parameters) to model all the well productivity index.

The sensitivity analysis of these AI-based predictive Shale Analytics models that will be explained in the following three steps includes about 250 wells. However, the XAI that will be provided in the next three steps can be applied to every single well that has been used to build the model using Artificial Intelligence. It should also be noted that the same types of XAI can also be applied to several wells and/or all the wells that have been used to develop this model. The sensitivity analyses that are demonstrated in the following parts of this chapter can also be applied to specific sectors of the reservoir (in the case of shale assets, they can be applied to each pad that includes a series of shale wells) that would include a certain number of wells and can be applied to all the wells in the entire field.

Step 1: Single Parameter Sensitivity Analysis

Total field measurements from 308 wells in Marcellus Shale were used for the development, validation, and AI explanation that are presented in this chapter. Data from about 250 wells in Marcellus Shale was used to develop this "AI-based Shale Predictive Analytics" model, and the remaining 58 wells were not used during the AI-based model development. Once the AI-based model development was completed using the 250 wells, in order to understand the explanation of the model behavior, the output of the model (6 Month Gas Production) was analyzed as a function of modifying every single input parameter to see if the results of such analyses made engineering (physics) sense. In other words, the idea is to be able to "EXPLAIN" the Artificial Intelligence model.

As it is shown in Figure 7.7, after the AI-based Shale Analytics model was completed through Training, Calibrations, and Validation using the 250 wells, the 58 wells that had not been used during the model development were used as Blind Validation to identify how accurately the AI-based Shale Analytics model can forecast the 6 Month Gas Production of the new wells. As shown in this figure, the R^2 of the 58 blind validation wells is 0.82. This would be the quality of this AI-based model to forecast the 6 Month Gas Production of all the new wells in Marcellus Shale. What is shown in this part of the chapter was developed using the "IMprove" software application from Intelligent Solutions, Inc.

Figures 7.7–7.15 show the Single Parameter Sensitivity Analysis of several wells. In these nine figures, the names of all the wells are provided on the top-left side, and the actual field measurement data that has been used to develop the AI-based engineering model from each well is provided on the bottom-left side. In Figure 7.7, it is shown that one of the specific wells (ISWN#032MSWPACW-3) is identified, and all the actual values of parameters from this specific well that were used during the building of this AI-based Shale Analytics model are also demonstrated (bottom-left-side). One of the input parameters that is selected is "porosity" in this figure.

Obviously, this particular well (as well as all other wells in this model) has a specific porosity value that has been measured and then used as part of the entire data set to build the AI-based predictive Shale Analytics model. The purely actual field measurement data model that has been developed for this field in Marcellus Shale has proven to provide good output (6 Month Gas Production) based on this specific value of porosity. The question is: what would happen to model output (6 Month Gas Production) as the value of the porosity for this specific well is modified? Keeping all other input parameters constant, the value of the porosity is changed (mentioned "Sweep" in this software) from minimum to maximum in 100 pieces. What would be the 6 Month Gas Production of the well for each new value of the porosity while all other parameters remain constant? (Figure 7.6).

Figure 7.7 shows the behavior of this AI-based Shale Analytics model of a specific well (ISWN#032MSWPACW-3). This "Predictive Shale Analytics" model explains that the hydrocarbon production of this particular well would have been higher if the porosity of the formation in this well was higher than what was actually measured.

FIGURE 7.6   The AI-based Shale Analytics model provided over 0.95% R^2 during training, calibration, and validation. Then the 58 blind validation wells that were not used during the model development show how accurately the AI-based Shale Analytics model can forecast the 6 Month Gas production of new wells.

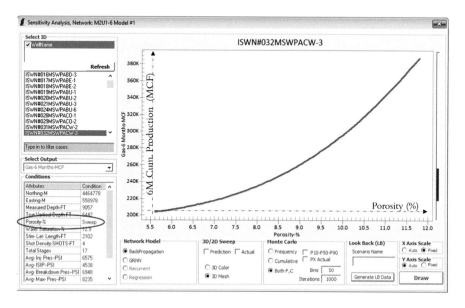

FIGURE 7.7   This figure shows that if all other variables are kept constant, then in ISWN#032MSWPACW-3 well increase in porosity will result in an increase in hydrocarbon production.

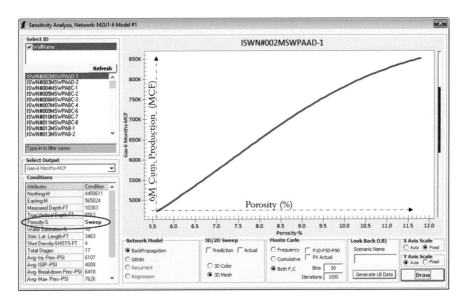

FIGURE 7.8   The same behavior in ISWN#002MSWPAAD-1 well (increase in porosity resulting in an increase in hydrocarbon production from wells) takes place for other wells in the same field.

From a petroleum engineering point of view, this makes perfect sense. This process can be repeated for every single well in this field. Examples of two more wells in this field are shown in Figure 7.10 (ISWN#002MSWPAAD-1) and Figure 7.9 (ISWN#012MSWPAB-1). It is important to note that while the production increases in each of these figures as a function of the enhancement of porosity, the behavior of the production in each of the wells is different from each other. It must be noted that all other field measurements for each of these wells are different from each other, and that is the reason for the production values for each porosity value. Both enhancements of production as a function of porosity and their amount and shape of difference that are not the same would have to do with other measured values. This would be an explanation for the AI-based Shale Analytics model. From a physics point of view, this also makes perfect sense.

What is very important to note is the fact that no specific mathematical equation (or pre-determined correlation through equation-based data) was used to develop this "AI-based Shale Analytics model". Figure 7.10 has performed Single Parameter Sensitivity Analysis on "Stimulated Lateral Length" on the well (ISWN#056MSWPADU-1), Figure 7.11 has done the same for the well (ISWN#032MSWPACW-3) and Figure 7.12 has done it for the well (ISWN#036MSWPACC-1). This same Single Parameter Sensitivity Analysis has been repeated for "Proppant per Stage" on Figure 7.13 for the well (ISWN#089MSWPAK-1), on Figure 7.14 for the well (ISWN#056MSWPADU-1), and finally on Figure 7.15 for the well (ISWN#076MSWPAHJ-2). The model behavior for all these wells and parameters makes perfect sense from a physics point of view. This is a good example of XAI in shale analytics.

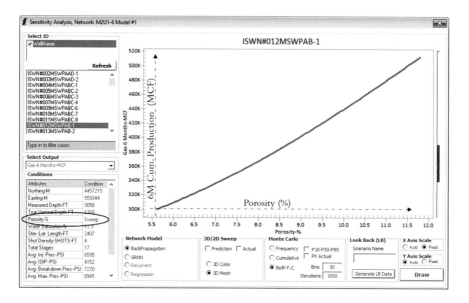

FIGURE 7.9   The same behavior in ISWN#012MSWPAB-1 well (increase in porosity resulting in an increase in hydrocarbon production from wells) takes place for other wells in the same field.

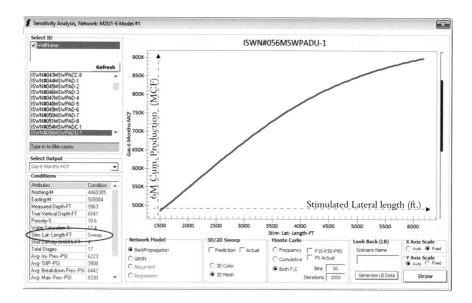

FIGURE 7.10   This figure shows that if all other variables are kept constant, then in ISWN#056MSWPADU-1 well increase in stimulated lateral length will result in an increase in hydrocarbon production.

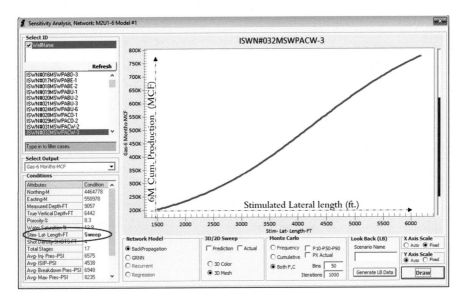

FIGURE 7.11 The same behavior in ISWN#032MSWPACW-3 well (increase in stimulated lateral length resulting in an increase in hydrocarbon production from wells) takes place for other wells in the same field.

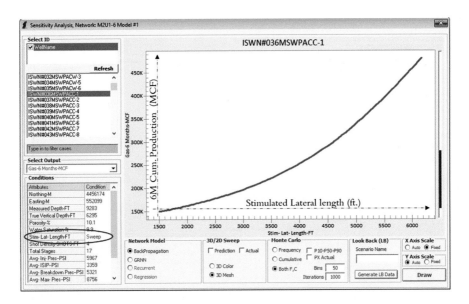

FIGURE 7.12 The same behavior in ISWN#036MSWPACC-1 well (increase in stimulated lateral length resulting in an increase in hydrocarbon production from wells) takes place for other wells in the same field.

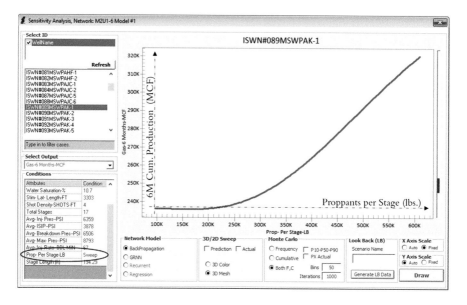

FIGURE 7.13 This figure shows that if all other variables are kept constant, then in ISWN#089MSWPAK-1 well increase in proppant per stage will result in an increase in hydrocarbon production.

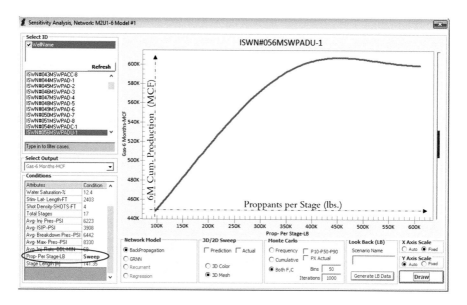

FIGURE 7.14 The same behavior in ISWN#056MSWPADU-1 well (increase in proppant per stage resulting in an increase in hydrocarbon production from wells) takes place for other wells in the same field.

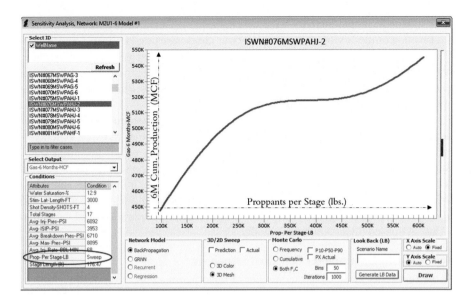

FIGURE 7.15   The same behavior in ISWN#076MSWPAHJ-2 well (increase in proppant per stage resulting in increase in hydrocarbon production from wells) takes place for other wells in the same field.

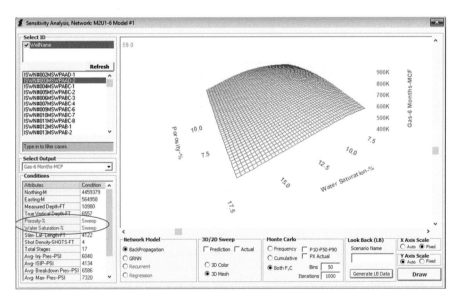

FIGURE 7.16   This figure shows the sensitivity of the hydrocarbon production from a given shale well as a function of two variables (porosity and initial water saturation) while all other variables are kept constant.

Step 2: Double Parameter Sensitivity Analysis

Analyzing the sensitivity of two parameters simultaneously, instead of a single parameter, will result in a three-dimensional figure. As shown in Figure 7.16, the modification of the "6 Month Gas productivity" (Z access in the plot) of an identified well is explained as a function of modifying of the two reservoir characteristics, "Porosity" and "Initial Water Saturation". While the Artificial Intelligence model has generated the productivity of this specific well, it is explaining how the well productivity may change as a function of modifications to these two reservoir characteristics while keeping all other input parameters constant.

It is interesting to note that the explanation of this "AI-based Shale Analytics" model makes perfect physics sense, since it is explaining that the well productivity will increase as the porosity of the reservoir increases while the initial water saturation decreases. Again, it is important to note that (a) the same type of general trend (increase of well productivity as porosity increases and initial water saturation decreases) takes place for every single well in this field, and (b) no mathematical equations were used to generate any data for the training, calibration, and validation of this "AI-based Shale Analytics" model.

Step 3: Multiple Parameter Sensitivity Analysis

As the sensitivity analysis of the number of input parameters increases to more than two parameters, providing figures with multiple dimensions will no longer be possible. The most common way of performing sensitivity analysis for multiple input parameters (more than two) is Monte Carlo Simulation (MCS). In MCS, there are no limitations on how many input parameters can be used in order to perform multiple-parameter sensitivity analysis.

The "AI-based Shale Analytics" model is used as the MCS's objective function. Once the number of parameters for the sensitivity analysis is identified, the process can be started. In the example shown in this part of the chapter in Figure 7.17, four parameters have been identified for sensitivity analysis. The four parameters that have been identified as shown in Figure 7.17 are porosity [%], initial water saturation [%], stimulated lateral length [ft.], and proppant per stage [lbs.]. Once the number of parameters for the sensitivity analyses has been identified, performing MSC requires the following three steps:

  a. Identification of the type of distribution for each parameter [uniform, triangular, Gaussian, … distribution],
  b. Number of times that the MCS's objective function must be executed (the "AI-based Shale Analytics" model in this case) to generate the results (model output – here: well productivity), and finally,
  c. Identification of the number of bins to demonstrate the distribution of the well productivity as a function modification of all the identified input parameters.

The above-mentioned three steps are identified in Figure 7.17. Results of the multiple parameter sensitivity analysis for three of the wells in this field are shown in Figure 7.17 (ISWN#002MSWPAAD-1), Figure 7.18 (ISWN#018MSWPABE-2), and Figure 7.19 (ISWN#018MSWPABE-2). In these analyses, four parameters (Porosity, Initial Water Saturation, Stimulated Lateral Length, and Proppant per Stage) were used to demonstrate the sensitivity of each well's productivity to their modifications, while all other input parameters for each well were kept constant. While the blue bar chart shows the distribution of the well productivity as a function of modification of these four parameters, the red curve shows the summation of the distribution.

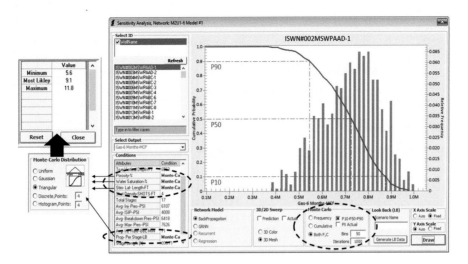

FIGURE 7.17   Sensitivity analysis of multiple parameters using MCS. The predictive model is deployed thousands of times in order to identify the potential distribution of hydrocarbon production as a function of changes in both reservoir characteristics (porosity and initial water saturation) and completion design (stimulated lateral length and proppant per stage).

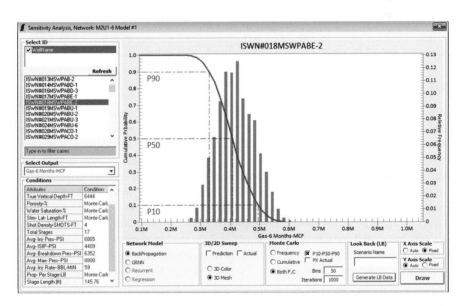

FIGURE 7.18   The AI model is deployed one thousand times for well ISWN#018MSWPABE-2 to identify the potential distribution of hydrocarbon production as a function of changes in both reservoir characteristics and completion design.

Using the red curve in these figures, P10 (almost the highest productivity of the well), P50 (almost the average productivity of the well), and P90 (almost the lowest productivity of the well) can be identified for each particular well as a function of modification of the parameters that are used for the sensitivity analysis. The characteristics

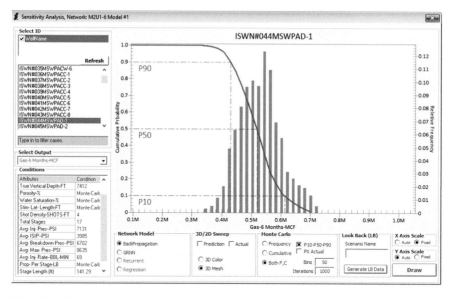

FIGURE 7.19 The AI model is deployed one thousand times for well ISWN#018MSWPABE-2 to identify the potential distribution of hydrocarbon production as a function of changes in both reservoir characteristics and completion design.

of the "blue bar chart" and the "red curve" explain the physics of the "AI-based Shale Analytics" model in many details for every single well in this field.

PART 3 OF XAI: TYPE CURVES

As was mentioned earlier in this chapter, Type Curves that are generated using mathematical equations are very "well-behaved" (continuous, non-linear, certain shape that changes in a similar fashion from curve to curve). Figure 7.20 demonstrates a few more examples of Type Curves that have been generated in reservoir engineering. The question is, "what is the main characteristic of a model that is capable of generating a series of well-behaved Type Curves?" The immediate, simple answer to this question would be: "The model that is capable of generating a series of well-behaved Type Curves is a physics-based model developed by one or more mathematical equations. The well-behaved Type Curves that clearly explain the behavior of the physics-based model are generated through the solutions of the mathematical equations".

The next question then would be: "What if it can be shown (and proven) that the model that has generated the well-behaved Type Curves has not been developed using any mathematical equations? What if this model is purely based on Artificial Intelligence?" Then the answer that would make sense would be:

A. The generation of these well-behaved Type Curves proves that the Artificial Intelligence that was used to build the purely data-driven model reasonably represents the physical phenomena that were being modeled,
B. The Artificial Intelligence predictive model can be clearly explained, and
C. The technique that was used to develop such a model must be a scientist and engineering solid technology.

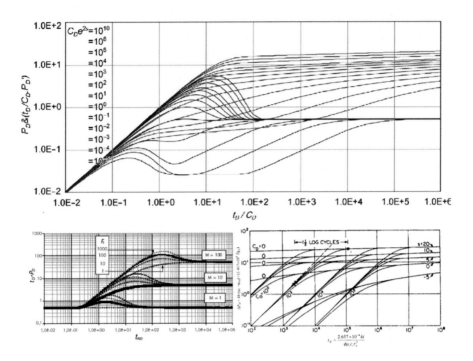

FIGURE 7.20   Type Curve examples of reservoir engineering.

XAI Model for Unconventional Reservoir – Shale Analytics

The main characteristics of the well-behaved Type Curves are the explainability of the model that has generated them. Figures 7.21 and 7.22 are good examples of "explaining" the model that has developed them. Figure 7.21 includes three series of graphs/ Type Curves. These Type Curves have "Stimulated Lateral Length (ft.)" as their x-axis and "6 Month Cumulative Gas Production (Mscf)" as their y-axis. In Figure 7.21, each of the curves in a graph represents the number of Hydraulic Fracturing Stages that are used during the completion of the well. There are six curves in each graph. Each of the graphs explains the behavior of a specific Marcellus Shale well in Southwestern Pennsylvania. The six curves in each graph show how each well's productivity changes as a function of the Number of Stages (from 15 to 40 stages) in Figure 7.21 and the "Stimulated Lateral Length (ft.)" in Figure 7.22. The three wells that are shown in Figures 7.21 and 7.22 belong to three different wells in three different locations in this specific Marcellus Shale asset.

Shale Analytics is the Science and Engineering Application of Artificial Intelligence in unconventional reservoirs. The Artificial Intelligence Predictive Analytics of Shale that are developed for completion and production optimization are purely based on actual field measurements. In Shale Analytics use and incorporation of any types of mathematical equations are avoided due to a lack of realistic understanding of the physics of fluid flow in shale plays and the shape and characteristics of the fractures that are created as a function of the implementation of hydraulic fracturing in natural fracture plays.

Once Artificial Intelligence Predictive Analytics of Shale is completed, through the generation of Type Curves, XAI is used in order to explain the physics of the completion and production of shale wells. Such explanations that are based on actual field measurements and completely avoid any kind of assumptions, interpretations, simplifications, and biases cannot be done through traditional approaches that have been used in the petroleum industry during the past decade. Traditional modeling approaches to hydrocarbon production from shale wells using Rate Transient Analysis (RTA) and Numerical Reservoir Simulation (NRS) include minimum amounts of actual field measurements and are fully controlled by soft (assumed) data. Soft (assumed) data that fully controls RTA and NRS include Fracture Half-Length, Fracture Height, Fracture Width, Fracture Conductivity, and even Stimulated Reservoir Volume (SRV), none of which is actually measured. Using these soft (assumed) data that can be generated by us and can have any values that we like them to have, instead of actual field measurements that are the main foundation of Shale Analytics, allows us to make any conclusions that we like to come up with even if they are 100% different from one another. In other words, such techniques can generate any kind of solution that we like to come up with and have absolutely nothing to do with facts or reality. It has to do with our objectives and has nothing to do with the realities associated with hydrocarbon production from shale wells.

Figures 7.21 and 7.22 show how XAI can provide information for every single well in a Marcellus Shale asset. Similar Type Curves can be generated for each pad or each well, any specific part of the shale asset, or for the entire shale asset. In Figure 7.21, the XAI shows that the 6 Month Cumulative Gas Production of Well #015 can increase from 850,000 MSCF to 1.3 million MSCF as the "Stimulated Lateral Length (ft.)" of this well increases from 3,500 to 10,000 ft. This increase in Gas Production as a function of Stimulated Lateral Length is non-linear and can change the way it is shown in this figure when the completion design of hydraulic fracturing changes from 15 to 40 stages as long as all other variables that have been used to build this model remain the same for this specific well. If the completion design includes 15 stages, then the Gas Production can increase from 850,000 MSCF to 1.05 million MSCF as the Stimulated Lateral Length increases from 3,500 to 10,000 ft.

If the completion design includes twice as many stages (30 stages), then the Gas Production can increase from 900,000 MSCF to 1.25 million MSCF as the Stimulated Lateral Length increases from 3,500 to 10,000 ft.

For Well #112 (the middle graph in Figure 7.21), the Type Curves explain that 6 Month Cumulative Gas Production can increase from 100,000 to 900,000 MSCF as the "Stimulated Lateral Length (ft.)" increases from 3,500 to 10,000 ft. This increase of Gas Production as a function of Stimulated Lateral Length is non-linear and can change the way it is shown in this figure when the completion design of hydraulic fracturing changes from 15 to 40 stages as long as all other variables that have been used to build this model remain the same for this specific well. If the completion design includes 15 stages, then the Gas Production can increase from 100,000 to 700,000 MSCF as the Stimulated Lateral Length increases from 3,500 to 10,000 ft. If the completion design includes twice as many stages (30 stages), then the Gas Production can increase from 100,000 to 820,000 MSCF as the Stimulated Lateral Length increases from 3,500 to 10,000 ft.

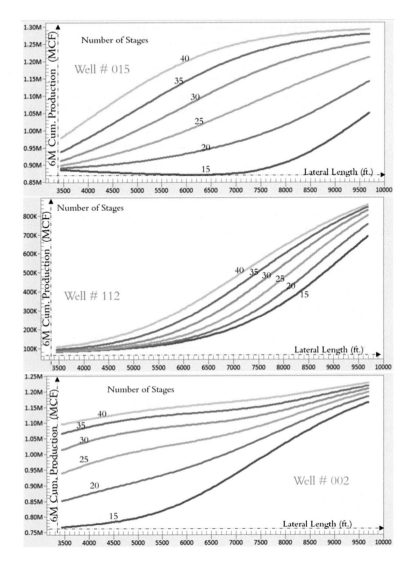

FIGURE 7.21   Well productivity as a function of stimulated lateral length and number of stage Type Curves.

For Well #002 (the bottom graph in Figure 7.21), the Type Curves explain that 6 Month Cumulative Gas Production can increases from 750,000 to 1.25 million MSCF as the "Stimulated Lateral Length (ft.)" of this well increases from 3,500 to 10,000 ft. This increase of Gas Production as a function of Stimulated Lateral Length is non-linear and can change the way it is shown in this figure when the completion design of hydraulic fracturing changes from 15 to 40 stages as long as all other variables that have been used to build this model remain the same for this specific well. If the completion design includes 15 stages, then the Gas Production can increase from 750,000 MSCF to 1.16 million MSCF as the Stimulated Lateral Length increases from 3,500 to 10,000 ft. If the completion design includes twice as many stages (30 stages), then the Gas Production can increase from 1.0 million MSCF to 1.21 million MSCF as the Stimulated Lateral Length increases from 3,500 to 10,000 ft (Figure 7.22).

Type Curves in Figure 7.22 explain that in this Marcellus Shale asset, 6 Month Cumulative Gas Production of Well #325 can increase from 0.1 million MSCF to 1.2 million MSCF as the "Stimulated Lateral Length (ft.)" of this well increases

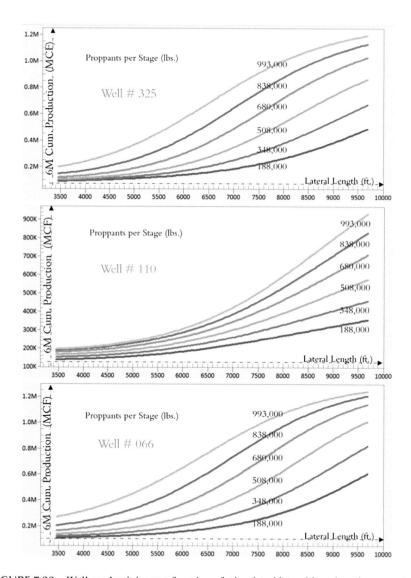

FIGURE 7.22 Well productivity as a function of stimulated lateral length and proppant per stage Type Curves.

from 3,500 to 10,000 ft. This increase of Gas Production as a function of Stimulated Lateral Length is non-linear and can change the way it is shown in this figure when the injection of proppant per stage in the completion design of hydraulic fracturing changes from 188,000 to 993,000 lbs. if all other variables that have been used to build this model remain the same for this specific well.

If the proppant per stage is 188,000 lbs., then the Gas Production can increase from 0.1 million MSCF to 0.5 million MSCF as the Stimulated Lateral Length increases from 3,500 to 10,000 ft. If the proppant per stage increases to 680,000 lbs., then the Gas Production can increase from 0.15 million MSCF to 1.03 million MSCF as the Stimulated Lateral Length increases from 3,500 to 10,000 ft.

For Well #110 (the middle graph in Figure 7.22) the Type Curves explain that 6 Month Cumulative Gas Production can increase from 100,000 to 950,000 MSCF as the "Stimulated Lateral Length (ft.)" of this well increases from 3,500 to 10,000 ft. This increase of Gas Production as a function of Stimulated Lateral Length is non-linear and can change the way it is shown in this figure when the proppant per stage change from 188,000 to 993,000 lbs. if all other variables that have been used to build this model remain the same for this specific well.

If the proppant per stage is 188,000 lbs., then the Gas Production can increase from 100,000 to 350,000 MSCF as the Stimulated Lateral Length increases from 3,500 to 10,000 ft. If the proppant per stage increases to 838,000 lbs., then the Gas Production can increase from 150,000 to 830,000 MSCF as the Stimulated Lateral Length increases from 3,500 to 10,000 ft.

For Well #066 (the bottom graph in Figure 7.22) the Type Curves explain that 6 Month Cumulative Gas Production can increase from 0.1 million MSCF to 1.22 million MSCF as the "Stimulated Lateral Length (ft.)" of this well increases from 3,500 to 10,000 ft. This increase of Gas Production as a function of Stimulated Lateral Length is non-linear and can change the way it is shown in this figure when the proppant per stage changes from 188,000 to 993,000 lbs. as long as all other variables that have been used to build this model remain the same for this specific well.

If the proppant per stage is 188,000 lbs., then the Gas Production can increase from 0.1 million to 0.6 million MSCF as the Stimulated Lateral Length increases from 3,500 to 10,000 ft. If the proppant per stage injection of the completion design increases to 993,000 lbs., then the Gas Production can increase from 0.25 million MSCF to 1.25 million MSCF as the Stimulated Lateral Length increases from 3,500 to 10,000 ft.

Repeating what was mentioned before, Shale Analytics (as well as all the applications of Artificial Intelligence in petroleum engineering) does not use any mathematical equations to generate data or to perform any kind of calculations to generate the types of Type Curves that are shown in Figures 7.21 and 7.22 Actual field measurements that are shown in Figure 7.23 were the source of the AI-based Shale Analytics that generate the Type Curves demonstrated in Figures 7.21 and 7.22. The complexity of the data shown in Figure 7.23 clarifies the quality of the XAI that is used in petroleum engineering.

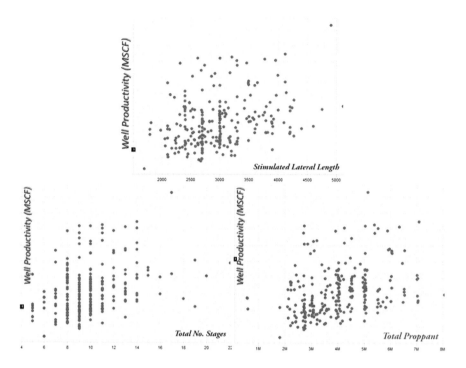

FIGURE 7.23   Actual field measurements from Marcellus Shale in Southwest Pennsylvania that were used during Artificial Intelligence Shale Predictive Analytics.

XAI Model for Conventional Reservoirs – Top-Down Modeling

Type Curves shown in Figures 7.24–7.26 were developed for a mature offshore field in northern Africa using Artificial Intelligence Reservoir Simulation and Modeling (Top-Down Modeling – TDM), which is a purely Data-Driven Reservoir Modeling technology [25]. These figures explain how and to what degree parameters such as Gas-Lift (Figure 7.24), Completion (Figure 7.25), and Porosity (Figure 7.26) can historically influence oil production for the entire field.

The well-behaved characteristics of these Type Curves that were developed using actual field measurements rather than mathematical equations clearly show how "XAI" has been used for the development of Artificial Intelligence Reservoir Simulation and Modeling (TDM). As was mentioned before, Artificial Intelligence Reservoir Simulation and Modeling (TDM) does not use any mathematical equations to generate dates or to perform any kind of calculations to generate the Type Curves that are shown in Figures 7.24–7.26. Actual field measurements that are shown in Figure 7.27 were the source of the Artificial Intelligence Reservoir Simulation and Modeling (TDM) that generates the Type Curves demonstrated in Figures 7.24–7.26. The complexity of the data shown in Figures 7.24–7.26 clarifies the quality of the XAI that is used in petroleum engineering.

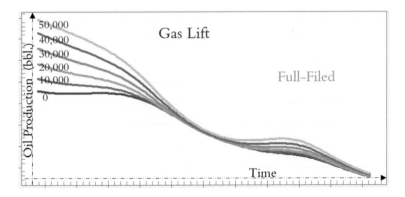

FIGURE 7.24 Gas-lift Type Curves of the well productivity for the entire field as a function of time (date).

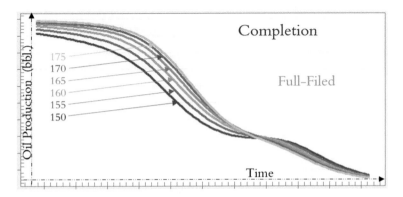

FIGURE 7.25 Completion Type Curves of the well productivity for the entire field as a function of time (date).

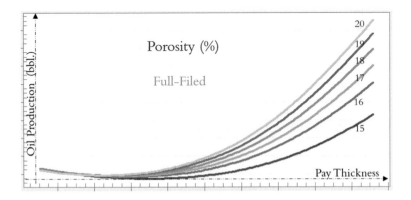

FIGURE 7.26 Porosity Type Curves of the well productivity for the entire field as a function of formation pay thickness.

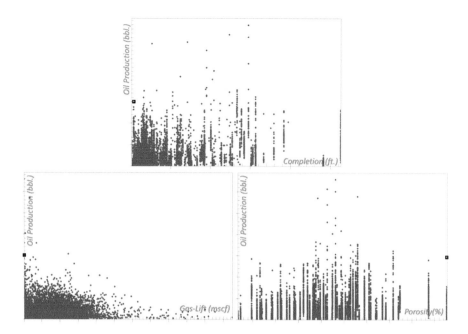

FIGURE 7.27   Actual field measurements from an off-shore mature field in Southeast Asia that was used during the artificial intelligence reservoir simulation and modeling (top-down modeling – TDM).

History of XAI in Petroleum Engineering

Since 2001, seven Society of Petroleum Engineering (SPE) papers have been published that include what is today called XAI. These seven SPE papers include a total of 38 figures that use the Artificial Intelligence Predictive Analytics models in order to explain how these models can explain the physical phenomena that were modeled using Artificial Intelligence purely based on actual field measurements rather than mathematical equations. To show the historical application of XAI in petroleum engineering, this section of the chapter shows most of the figures from these seven SPE technical papers.

The first technical paper that included the first version of what is today called XAI was an SPE paper (SPE 72385) that was published in 2001 [26]. This paper is about the application of Artificial Intelligence and Machine Learning to modeling hydraulic fractures. Figure 7.28 shows a couple of XAI-related figures from this SPE paper. In this paper, the Artificial Intelligence model that was developed (trained, calibrated, and validated) using an artificial neural network was used to explain the impact of three hydraulic fracturing-related parameters.

These hydraulic fracturing-related parameters are the number of perforations, injection rate, and total amount of water that is injected. The XAI in this paper explains the impact of these parameters on the well productivity (5-Year Cumulative Gas Production). Such explanations of the influence of these parameters on well productivity are used to identify well productivity optimization as a function of the

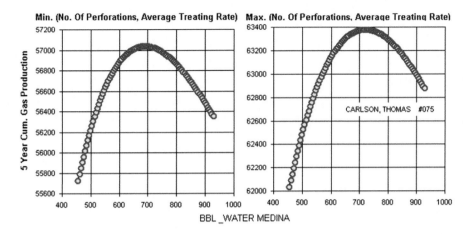

FIGURE 7.28   SPE 72385 paper's Figure 7. Single well "Carlson, Thomas #075" analysis – Sensitivity to the amount of water injected.

number of perforations, injection rate, and total amount of water that is injected. Figure 7.28, which shows Figure 7 of this paper, explains the conditions under which the well productivity (5-Year Cumulative Gas Production) can reach its maximum value. For this particular well (Carlson, Thomas #075), the 5-Year Cumulative Gas Production can reach its maximum value of 57,050 MSCF as long as the number of perforations and the rate of injection are kept at their minimum values (injection rate at 25 bbl./min and 8 perforations) and can reach its maximum value of 63,400 MSCF as long as the number of perforations and the rate of injection are kept at their maximum values (60 bbl./min for injection rate and 26 perforations).

As shown in Figure 7.28, while these parameters (the number of perforations and the rate of injection) are kept at their minimum value, the highest 5-Year Cumulative Gas Production (57,050 MSCF) is achieved at 680 barrels of water. On the other hand, when these values are kept at their maximum, the highest 5-Year Cumulative Gas Production (63,400 MSCF) is achieved at about 725 barrels of water injected. The conclusion may be that for this well, the ideal water injection volume is about 700 barrels. This SPE paper includes two more similar figures showing an explanation for two more wells in the same field (Figure 7.29).

Similar to the paper that was published in 2001, the second technical paper that included XAI was an SPE paper (SPE 77597) that was presented at ATCE and was published in 2002 [27]. This paper is also about the application of AI and Machine Learning in modeling re-fracturing (re-stimulation) in DJ-Basin, Colorado. Figure 7.29 is an explanation of the modification of the well productivity as a function of parameters such as the amount of injected proppant (sand) and fluid as well as the number of perforations. In this figure, the monthly barrel of oil equivalent as well as productivity are shown for Bohlender 8-5 wells during the re-fracturing process.

The third technical paper that included XAI was an SPE paper (SPE 77659) that was also published in 2002 [28]. This paper covers purely data-driven modeling of BP's Prudhoe Bay (Alaska, USA) surface facility using Artificial Intelligence and

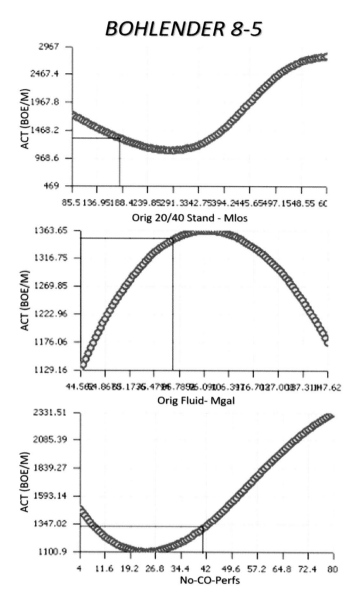

FIGURE 7.29 SPE 77597 paper's Figure 8. Sensitivity of the post-restimulation actual peak (after an average restimulation job) to different values of three parameters being studied, namely, amount of sand, amount of fluid, and number of perforations for the well Bohlender 8-5.

Machine Learning. This paper includes figures that generate XAI covering separator pressure, temperature, hydrocarbon rates, and compressor inlet suction pressure of eight different three-phase separator facilities. Figures 7.30 and 7.31 show two such figures from this SPE paper.

The fourth technical paper that included XAI was an SPE paper (SPE 89033) that was published in 2004 [29] and included more examples of XAI from Prudhoe Bay's

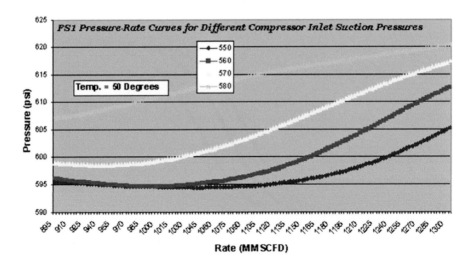

FIGURE 7.30 SPE 77659 paper's Figure 13. Rate vs. pressure curves for FS1 (three-phase separator facility) at 50 degrees of temperature and different compressor inlet suction pressures.

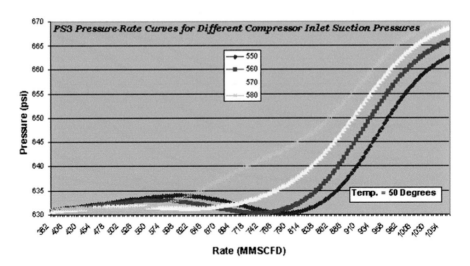

FIGURE 7.31 SPE 77659 paper's Figure 15. Rate vs. pressure curves for FS3 (three-phase separator facility) at 50 degrees of temperature and different compressor.

surface facility's purely data-driven model. This paper includes seven figures that cover XAI of Prudhoe Bay's surface facility. Figures 7.32 and 7.33 show two such figures from this SPE paper.

The fifth technical paper that included XAI was an SPE paper (SPE 95942) that was published in 2005 [30]. This paper covers XAI of hydraulic fracturing in the Golden Trend filed of Oklahoma. This paper includes two figures that show XAI results for the hydraulic fracturing of multiple individual wells in this field.

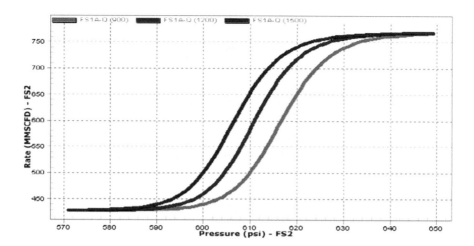

FIGURE 7.32   SPE 89033 paper's Figure 12. FS2 (3-phase separator facility) rate behavior as a function of pressure @ FS2 and FS1A rates.

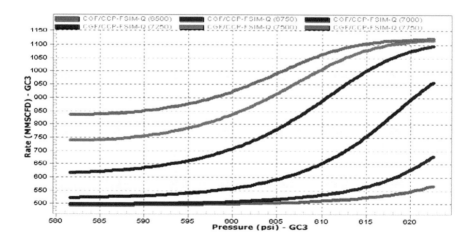

FIGURE 7.33   SPE 89033 paper's Figure 19. GC3 (3-phase separator facility) rate behavior as a function of pressure @ GC3 and CCP rates.

Figure 7.34 explains how shot per foot (x-axis) and injection rates – BPM/ft. (y-axis) influence 30-Year EUR well productivity, for three individual wells. These wells show different types of responses as the number of perforations and average rate of injection per foot of pay thickness change. The production response is different for each of these wells as the number of perforations and the average injection rates start to increase.

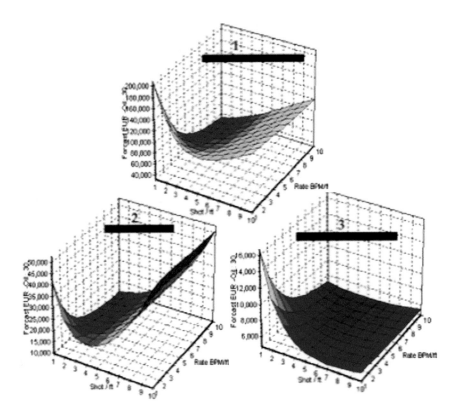

FIGURE 7.34 SPE 95942 paper's Figure 13. Sensitivity analysis for "Shot/ft" and "Rate" for three wells in the database.

The sixth technical paper that included XAI was an SPE paper (SPE 101474) that was published in 2006 [31]. This paper covers XAI of Smart Proxy Modeling of traditional numerical simulation models. Figure 7.35 provides Type Curves of the model that demonstrate the impact of a specific reservoir parameter on oil production, while Figure 7.36 provides Type Curves of the model demonstrating the impact of another specific reservoir parameter on water cut. The seventh technical paper that included XAI was an SPE paper (SPE 139032) that was published in 2010 [32]. This paper covers the completion design characteristics of hydraulic fracturing in Bakken Shale. The XAI model that is shown in this paper was developed using "IMagine" software application that is used to develop TDM [33].

Figure 7.37 demonstrates the use of XAI to explain the oil production characteristics as a function of "Lateral Length" and "Injected Fracturing Fluid" (graph on the left) and the oil production characteristics as a function of "Pay Thickness" and "Lateral Length" (graph on the right).

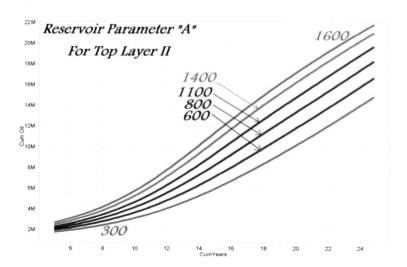

FIGURE 7.35   SPE 101474 paper's Figure 10. Behavior of 5-Year Cumulative Gas Production as a function of time for different values of parameter "A" of top layer II. This can be considered a type curve for this reservoir.

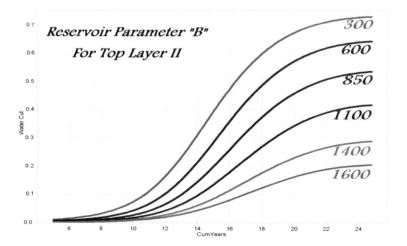

FIGURE 7.36   SPE 101474 paper's Figure 11. Behavior of instantaneous water cut as a function of time for different values of parameter "B" of top layer II. This can be considered a type curve for this reservoir.

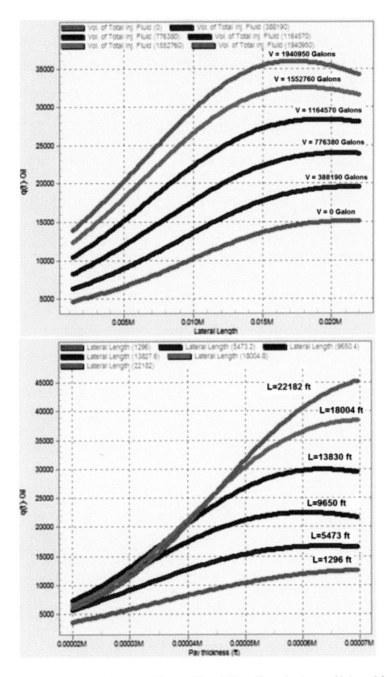

FIGURE 7.37   SPE 139032 paper's Figures 27 and 28 – effect of volume of injected fracturing fluid and lateral length on production rate (Middle Bakken model) – effect of pay thickness and lateral length on production rate (Middle Bakken model).

NOTES

1 http://IntelligentSolutionsInc.com
2 https://shahab-mohaghegh.medium.com/petroleum-data-analytics-frequently-asked-questions-513090783bcf

References

1. Minsky, M. and Papert, S., *Perceptrons: An Introduction to Computational Geometry*, The MIT Press, Cambridge, MA; London, 1969, ISBN: 0 262 13043 2. https://www.amazon. com/Perceptrons-MIT-Press-Introduction-Computational/dp/0262534770/ref=asc_df_0 262534770/?tag=hyprod-20&linkCode=df0&hvadid=312060980065&hvpos=&hvnetw =g&hvrand=14936349446638093033&hvpone=&hvptwo=&hvqmt=&hvdev=c&hvdvc mdl=&hvlocint=&hvlocphy=9009438&hvtargid=pla-466303118778&psc=1

2. Hodges, A., *A Short Biography of Alan Turing*. https://www.turing.org.uk/publications/ dnb.html

3. Mohaghegh, S.D., "Virtual-Intelligence Applications in Petroleum Engineering: Part 1 - Artificial Neural Networks" *Journal of Petroleum Engineering (JPT)*, 2000, 64–73.

4. Mohaghegh, S.D, "Virtual-Intelligence Applications in Petroleum Engineering: Part 2 - Evolutionary Computing" *Journal of Petroleum Engineering (JPT)*, 2000, 40–46

5. Mohaghegh, S.D. "Virtual-Intelligence Applications in Petroleum Engineering: Part 3 - Fuzzy Logic" *Journal of Petroleum Engineering (JPT)*, 2000, 82–87.

6. Harari, Y.N. *Sapiens: A Brief History of Humankind*. 2011. *https://www.amazon.com/ Sapiens-Humankind-Yuval-Noah-Harari/dp/0062316095*

7. Darwin, C. *The Descent of Man, and Selection in Relation to Sex*, 1871. *https://www. google.com/books/edition/The_Descent_of_Man_and_Selection_in_Rela/gL3jkKUq lDMC?hl=en&gbpv=1&printsec=frontcover*

8. CRISPR. Gene-Bases Medicines. Transforming the Lives of Patients with Serious Diseases. https://crisprtx.com/

9. *News Releases: Hitachi and EERC Propose Innovative Solutions for Optimizing.* Published by The Bakken, March 3, 2014. *https://www.hitachi.us/press/03312014*

10. SPE Petro Talk Video on YouTube. https://www.youtube.com/watch?v=7LvblCJYSV0

11. Artificial Intelligence. *Stanford Encyclopedia of Philosophy*. First published, Thu, July 12, 2018. *https://www.hitachi.us/press/03312014*

12. Breiman, L.. "Statistical Modeling: The Two Cultures" *Statistical Science*, 2001, Vol. 16. No. 3, 199–231. *https://www2.math.uu.se/~thulin/mm/breiman.pdf.*

13. Nisbet, R., Elder, J., and Miner, G., *Handbook of Statistical Analysis & Data Mining Applications*. Published by Elsevier, 2009. *https://www.google.com/books/edition/ Handbook_of_Statistical_Analysis_and_Dat/U5np34a5fmQC?hl=en&gbpv=1&printse c=frontcover*

14. Abu-Mostafe, Y., *The Learning Problem Lecture*. Published by Caltech (The California Institute of Technology in Pasadena, California). https://www.youtube.com/ watch?v=mbyG85GZ0PI

15. Crawford, K. and Paglen, T., *Excavating AI, the Politics of Images in Machine Learning Training Sets*, September 2019. https://www.excavating.ai/

16. Lim, H. "7 Types of Data Bias in Machine Learning", July 20, 2020. *https://lionbridge. ai/articles/7-types-of-data-bias-in-machine-learning/*

17. Doshi, T., "Introducing the Inclusive Images Competition", September 6, 2018, Product Manager, Google AI - Google AI Blog. *https://ai.googleblog.com/2018/09/introduc- ing-inclusive-images-competition.html*

18. Dave, P., "Exclusive Google Searches for New Measure of Skin Tones to Curb Bias in Products", June 2021. *https://www.reuters.com/business/sustainable-business/exclusive-google-searc hes-new-measure-skin-tones-curb-bias-products-2021-06-18/?utm_medium=techboard. fri.media.20210618&utm_source=email&utm_content=&utm_campaign=campaign*

19. Dastin, J., *Amazon Scraps Secret AI Recruiting Tool That Showed Bias Against Women*. Published by Reuters - San Francisco, October 2018. *https://www.reuters.com/article/us-amazon-com-jobs-automation-insight/amazon-scraps-secret-ai-recruiting-tool-that-showed-bias-against-women-idUSKCN1MK08G*

20. Hao, K., *Making Face Recognition Less Biased Doesn't Make It Less Scary*. Published by MIT Technology Review. Tech Policy - AI Ethics. https://www.technologyreview.com/2019/01/29/137676/making-face-recognition-less-biased-doesnt-make-it-less-scary/

21. Hao, K., This is How AI Bias Really Happens - And Why It's So Hard to Fix. Published by MIT Technology Review. Tech Policy - AI Ethics. https://www.technologyreview.com/2019/02/04/137602/this-is-how-ai-bias-really-happensand-why-its-so-hard-to-fix/

22. Terence, S., Real-Life Examples of Discriminating Artificial Intelligence. Real-Life Examples of AI Algorithms Demonstrating Bias and Prejudice. June 2020. https://towardsdatascience.com/real-life-examples-of-discriminating-artificial-intelligence-cae395a90070Sdfg

23. Schmelzer, R., Understanding Explainable AI, Published by Forbes. July 23, 2019. *https://www.forbes.com/sites/cognitiveworld/2019/07/23/understanding-explainable-ai/?sh=1f83c5ce7c9e*

24. Mohaghegh, S.D., *Shale Analytics - Data-Driven Analytics in Unconventional Resources*. Published by Springer International, 2017. Hard Cover print (ISBN: 978-3-319-48751-9). https://www.springer.com/gp/book/9783319487519

25. Mohaghegh, S.D., "Data-Driven Reservoir Modeling". Published by Society of Petroleum Engineering (SPE), 2018. Hard Cover print (ISBN: 978-1-61399-560-0). https://store.spe.org/DATA-DRIVEN-RESERVOIR-MODELING-P1054.ASPX

26. Mohaghegh, S.D., "Identifying Best Practices in Hydraulic Fracturing Using Virtual Intelligence Techniques". SPE 72385, SPE Eastern Regional Meeting held in Canton, Ohio, October 17–19, 2001.

27. Mohaghegh, S.D., "Identification of Successful Practices in Hydraulic Fracturing Using Intelligent Data Mining Tools; Application to the Codell Formation in the DJ -Basin". SPE 77597, SPE Annual Technical Conference and Exhibition held in San Antonio, Texas, September 29–October 2, 2002.

28. Mohaghegh, S.D., Hutchins, L. and Sisk, C., "Prudhoe Bay Oil Production Optimization: Using Virtual Intelligence Techniques, Stage One: Neural Model Building". SPE 77659, SPE Annual Technical Conference and Exhibition held in San Antonio, Texas, September, 29–October 2, 2002.

29. Mohaghegh, S.D., "Recent Development in Application of Artificial Intelligence in Petroleum Engineering". Published by West Virginia University & Intelligent Solutions, Inc. SPE 89033. JPT Distinguished Author Series, April 2004.

30. Mohaghegh, S.D., Gaskari, R., Popa, A., Salehi, I. and Ameri, S., "Analysis of Best Hydraulic Fracturing Practices in the Golden Trend Fields of Oklahoma". SPE 95942. SPE Annual Technical Conference and Exhibition held in Dallas, Texas, U.S.A., October 9–12, 2005.

31. Mohaghegh, S.D., Hafez H., Razi G., Masoud H., and Maher K., "Uncertainty Analysis of a Giant Oil Field in the Middle East Using Surrogate Reservoir Model". SPE 101474. Abu Dhabi International Petroleum Exhibition and Conference held in Abu Dhabi, U.A.E., November 5–8, 2006.

32. Mohaghegh, S.D., "Field Development Strategies for Bakken Shale Formation". SPE-139032. SPE Eastern Regional Meeting held in Morgantown, West Virginia, USA, October 12–14, 2010.

33. Intelligent Solutions, Inc. (ISI). https://www.IntelligenetSolutionsInc.com

Index